Gravitational Lensing and Optical Geometry

Gravitational Lensing and Optical Geometry: A Centennial Perspective

Editor

Marcus C. Werner

MDPI • Basel • Beijing • Wuhan • Barcelona • Belgrade • Manchester • Tokyo • Cluj • Tianjin

Editor
Marcus C. Werner
Duke Kunshan University
China

Editorial Office
MDPI
St. Alban-Anlage 66
4052 Basel, Switzerland

This is a reprint of articles from the Special Issue published online in the open access journal *Universe* (ISSN 2218-1997) (available at: https://www.mdpi.com/journal/universe/special_issues/gravitational_lensing_optical_geometry).

For citation purposes, cite each article independently as indicated on the article page online and as indicated below:

LastName, A.A.; LastName, B.B.; LastName, C.C. Article Title. *Journal Name* **Year**, *Article Number*, Page Range.

ISBN 978-3-03943-286-8 (Hbk)
ISBN 978-3-03943-287-5 (PDF)

© 2020 by the authors. Articles in this book are Open Access and distributed under the Creative Commons Attribution (CC BY) license, which allows users to download, copy and build upon published articles, as long as the author and publisher are properly credited, which ensures maximum dissemination and a wider impact of our publications.

The book as a whole is distributed by MDPI under the terms and conditions of the Creative Commons license CC BY-NC-ND.

Contents

About the Editor vii

Preface to "Gravitational Lensing and Optical Geometry: A Centennial Perspective" ix

Amir B. Aazami, Charles R. Keeton and Arlie O. Petters
Magnification Cross Sections for the Elliptic Umbilic Caustic Surface
Reprinted from: *Universe* **2019**, *5*, 161, doi:10.3390/universe5070161 1

Valerio Bozza, Silvia Pietroni and Chiara Melchiorre
Caustics in Gravitational Lensing by Mixed Binary Systems
Reprinted from: *Universe* **2020**, *6*, 106, doi:10.3390/universe6080106 9

Toshiaki Ono and Hideki Asada
The Effects of Finite Distance on the GravitationalDeflection Angle of Light
Reprinted from: *Universe* **2019**, *5*, 218, doi:10.3390/universe5110218 31

Ali Övgün
Deflection Angle of Photons through Dark Matter by Black Holes and Wormholes Using Gauss–Bonnet Theorem
Reprinted from: *Universe* **2019**, *5*, 115, doi:10.3390/universe5050115 75

Sumarna Haroon, Kimet Jusufi and Mubasher Jamil
Shadow Images of a Rotating Dyonic Black Hole with a Global Monopole Surrounded by Perfect Fluid
Reprinted from: **2020**, *6*, 23, doi:10.3390/universe6020023 85

Pedro V. P. Cunha, Carlos A. R. Herdeiro and Eugen Radu
EHT Constraint on the Ultralight Scalar Hair
of the M87 Supermassive Black Hole
Reprinted from: *Universe* **2019**, *5*, 220, doi:10.3390/universe5120220 103

About the Editor

Marcus C. Werner (Prof. Dr.) After obtaining his PhD at Cambridge University in 2009, he taught at Duke University's Mathematics Department. He then moved to Japan, where he was a researcher at the Kavli Institute for the Physics and Mathematics of the Universe, Tokyo University, before becoming a Hakubi assistant professor at Kyoto University. He joined Duke Kunshan University in 2020 as an associate professor of applied mathematics.

Preface to "Gravitational Lensing and Optical Geometry: A Centennial Perspective"

Gravitational lensing is the deflection of light under the influence of gravity as predicted by Einstein's general relativity . In 1919, Eddington's expeditions to Principe and Brazil to observe the gravitational deflection of star light during a total solar eclipse led to the corroboration of General Relativity.

To mark the centennial of this important discovery, a Special Issue of Universe entitled 'Gravitational Lensing and Optical Geometry: A Centennial Perspective' was dedicated to the theoretical aspects of gravitational lensing.

This has become a thriving subject at the interface of mathematics, astronomy and theoretical physics, and the variety of affiliations of the contributing authors testifies to that. Additionally, several mathematical approaches have been developed to study this effect, and one can distinguish three broad categories: firstly, the standard thin lens approximation in 3-space, which is often used in astronomy and has been found to be of great mathematical interest, for instance with its applications of singularity theory to investigate caustics; secondly, a differential geometrical formalism in 3-space, such as optical geometry and the related approaches: in this setting, the Gauss–Bonnet theorem has proven useful for finding the deflection angle and topological criteria of image multiplicity; and thirdly, the study of null geodesics in 4-dimensional spacetime of general relativity and its modifications. The recent observation of the so-called black hole shadow of M87* by the Event Horizon Telescope has made black hole photon regions a subject of great current interest as well.

The articles assembled in this volume reflect this range of topics and have been arranged thematically along these lines. I would also like to take this opportunity to thank the authors again for contributing to this Special Issue commemorating the foundational event of gravitational lensing studies.

Marcus C. Werner
Editor

Article

Magnification Cross Sections for the Elliptic Umbilic Caustic Surface

Amir B. Aazami [1,*], Charles R. Keeton [2] and Arlie O. Petters [3]

1. Department of Mathematics and Computer Science, Clark University, Worcester, MA 01610, USA
2. Department of Physics and Astronomy, Rutgers University, 136 Frelinghuysen Road, Piscataway, NJ 08854-8019, USA
3. Departments of Mathematics and Physics, Duke University, Science Drive, Durham, NC 27708-0320, USA
* Correspondence: aaazami@clarku.edu

Received: 29 May 2019; Accepted: 25 June 2019; Published: 2 July 2019

Abstract: In gravitational lensing, magnification cross sections characterize the probability that a light source will have magnification greater than some fixed value, which is useful in a variety of applications. The (area) cross section is known to scale as μ^{-2} for fold caustics and $\mu^{-2.5}$ for cusp caustics. We aim to extend the results to higher-order caustic singularities, focusing on the elliptic umbilic, which can be manifested in lensing systems with two or three galaxies. The elliptic umbilic has a caustic surface, and we show that the volume cross section scales as $\mu^{-2.5}$ in the two-image region and μ^{-2} in the four-image region, where μ is the total unsigned magnification. In both cases our results are supported both numerically and analytically.

Keywords: strong gravitational lensing; magnification cross sections; caustics

1. Introduction

Magnification cross sections are an important tool in gravitational lensing. Knowing the magnification cross section allows one to determine the probability that a light source will have magnification greater than some fixed value. This in turn gives information about the accuracy of cosmological models, since these predict different probabilities regarding source magnifications (see Schneider et al. 1992 [1], Kaiser 1992 [2], Bartelmann et al. 1998 [3], and Petters et al. 2001 ([4] Chapter 13)). Knowing magnification cross sections is also important for observational programs that use lensing magnification to help detect extremely faint galaxies (see, e.g., Lotz et al. 2017 [5], and references therein).

It is well known that for the so-called "fold" and "cusp" caustic singularities, the area cross sections scale asymptotically as μ^{-2} and $\mu^{-2.5}$, respectively, where μ is the total unsigned magnification of a lensed source (for the cusp caustic, it is assumed that the source lies in the one-image region locally). In the case of single-plane lensing, the fold scaling was determined by Blandford and Narayan 1986 [6], while the cusp scaling was determined by Mao 1992 [7] and Schneider and Weiss 1992 [8]. For multiple-plane lensing, the fold and cusp scalings were determined by Petters et al. ([4] Chapter 13).

In this paper we commence the study of magnification cross sections of "higher-order" caustic surfaces; in particular, we derive the (single-plane) asymptotic limit of the magnification cross section for the "elliptic umbilic" caustic surface, which is not a curve but rather a two-dimensional stable caustic surface in a three-dimensional parameter space (for precise definitions, consult, e.g., Arnold 1973 [9], Callahan 1974 & 1977 [10,11], Majthay 1985 [12], Arnold et al. 1985 [13], Petters 1993 [14], [1,4]); its magnification "cross section" is therefore a region with volume. As hypothesized in Rusin et al. 2001 [15] and Blandford 2001 [16], the elliptic umbilic is likely to be manifested inside a triangle formed by three lensing galaxies, and to involve one positive-parity and three negative-parity images. As shown in Shin and Evans 2008 [17] and de Xivry and Marshall 2009 [18], elliptic umbilics can also appear in lensing by binary galaxies, if the binary separation is small enough. The asymptotic scaling of the elliptic umbilic volume cross section will depend on whether the source gives rise to two or four lensed images locally. We show that, in the two-image region, this volume cross section scales to leading order as $\mu^{-2.5}$, whereas in the four-image region, its leading order scales as μ^{-2}. In both cases, our results are supported numerically and analytically. In our derivation of μ^{-2} in the four-image region, we make use of a certain magnification relation that holds for higher-order caustic singularities, and the elliptic umbilic in particular, that was shown to hold in Aazami and Petters 2009 [19].

2. The Elliptic Umbilic Caustic Surface

Let (x,y) denote coordinates on the lens plane and (s_1,s_2) coordinates on the source plane. Like all higher-order caustic surfaces, the "elliptic umbilic caustic" has parameters in addition to the two source plane coordinates. (Such higher-order parameters can, depending on the setting, be used to model the source redshift, radii of galaxies, ellipticities, distance along the line of sight, etc., of the lens system in question; see, e.g., ([1] Chapter 8)). A gravitational lensing map in the neighborhood of an elliptic umbilic critical point, as derived in ([1] Chapter 5), takes the form

$$\eta(x,y) = (x^2 - y^2, -2xy + 4cy) = (s_1, s_2), \qquad (1)$$

where $c \in \mathbb{R}$ is a parameter in addition to the source plane coordinates s_1, s_2; i.e., the elliptic umbilic caustic is a surface in the parameter space $\{(s_1, s_2, c)\} = \mathbb{R}^3$, not a curve in the source plane $S := \{(s_1, s_2)\} = \mathbb{R}^2$. Accordingly, the "magnification cross section" is a (three-dimensional) volume, and the asymptotic scaling is the leading order term in the limit as the magnification goes to infinity. Figure 1 shows the elliptic umbilic caustic surface, while Figure 2 shows a c-slice of it on the source plane S.

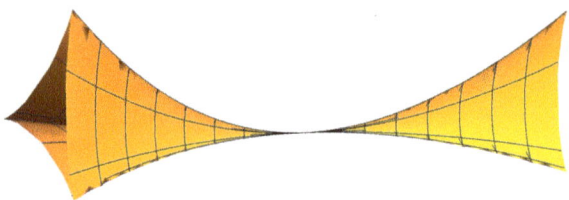

Figure 1. The elliptic umbilic caustic surface in three-dimensional parameter space $\{(s_1, s_2, c)\} = \mathbb{R}^3$; s_1, s_2 are source plane coordinates and $c \in \mathbb{R}$ an additional parameter. When $c = 0$ the elliptic umbilic is the central point shown. A c-slice of this caustic is shown in Figure 2.

If a source located at (s_1, s_2) on the source plane has a lensed image located at (x,y) on the lens plane, then the magnification $\tilde{\mu}$ of this lensed image is given by

$$\tilde{\mu}(x,y) = \frac{1}{\det(\mathrm{Jac}\,\eta)(x,y)} = \frac{1}{8cx - 4(x^2 + y^2)}. \qquad (2)$$

Now fix $c \in \mathbb{R}$ and $\mu > 0$. Consider first the four-image region enclosed by the caustic curve in Figure 2. Let C_μ denote the subset consisting of those source positions (s_1, s_2) in the four-image region with total unsigned magnification equal to μ, where the unsigned magnification of each image (x_i, y_i) belonging to (s_1, s_2) is given by $|\tilde{\mu}(x_i, y_i)|$ in Equation (2). C_μ will consist of the closed dashed curve inside the caustic curve in Figure 2; any source inside the dashed region will have total unsigned magnification less than μ, while a source in the region D will have total unsigned magnification equal to μ. Likewise for the two-image region, which is the (unbounded) region outside the caustic curve: For

μ large enough, the enclosed regions A, B, and C comprise those sources whose two lensed images will have total unsigned magnification greater than μ. (By symmetry, A, B, and C all have the same areas).

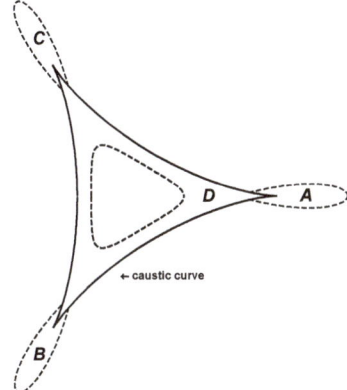

Figure 2. A generic c-slice of the elliptic umbilic caustic surface, which is the triangular-shaped solid curve. For level curves in the two-image region, which is the (unbounded) region outside the closed caustic curve, the magnification volume cross section $\sigma(\mu)$ is, for μ large enough, the sum of the closed regions A, B, C. A source inside regions A, B, or C has two lensed images and total unsigned magnification greater than μ. For level curves in the four-image region, which is inside the triangular-shaped solid curve, $\sigma(\mu)$ is, for μ large enough, the region D inside the triangular-shaped caustic but outside the closed dashed curve within. A source in region D has four lensed images and total unsigned magnification greater than μ. Note that the actual caustic and level "curve" are both surfaces in $\mathbb{R}^3$; what is shown here is a c-slice of this surface on the source plane.

Whether in the two- or the four-image region, the asymptotic scaling there is determined as follows. The areas labeled A, B, C, and D will in general be functions of μ and c; denote any one of these, e.g., by $A(\mu, c)$. To obtain a volume section, denoted $V(\mu)$, we integrate

$$V(\mu) = \int_{c_1}^{c_2} A(\mu, c) \, dc, \tag{3}$$

where $c_1 < c_2$ are arbitrary but of the same sign. Finally, we take the limit $\lim_{\mu \to \infty} V(\mu)$ and identify the leading order term; this is the asymptotic scaling of the elliptic umbilic magnification volume.

We first examine the scalings numerically. For a given c-slice, we compute the total magnification on a grid in plane S, as shown in Figure 3. The grid is adaptive, meaning that more grid points are used in regions where the magnification changes quickly and high resolution is needed to obtain accurate results. We use the grid to approximate the area integral and compute $A(\mu, c)$. We then combine different c-slices to approximate the integral in Equation (3) and obtain $V(\mu)$. Figure 4 shows examples of $A(\mu, c)$ and $V(\mu)$ for the two-image region, while Figure 5 shows $A(\mu, c)$ for the four-image region.

We support our numerical findings with the following analytical arguments. For the four-image region, there is in fact a succinct argument that confirms the numerical result μ^{-2}, as follows. For any source in the four-image region (and assuming $c \neq 0$), the lensing map of Equation (1) has three lensed images with negative magnification and one lensed image with positive magnification, where the magnification of a lensed image is given by Equation (2). Fix $\tilde{\mu} > 0$ and let $S_{\tilde{\mu}}$ denote the set of sources in the four-image region whose one positive-magnification image has magnification $\tilde{\mu}$; as usual, let μ denote the total unsigned magnification of this source. In fact $S_{\tilde{\mu}}$ comprises the closed dashed curve inside the caustic that we saw in Figure 2. It is straightforward to compute that the area labeled D—that is, the area outside $S_{\tilde{\mu}}$ but inside the caustic—is equal to

$$2\pi c^4 - \left(2\pi c^4 - \frac{\pi}{8\tilde{\mu}^2}\right) = \frac{\pi}{8\tilde{\mu}^2}. \qquad (4)$$

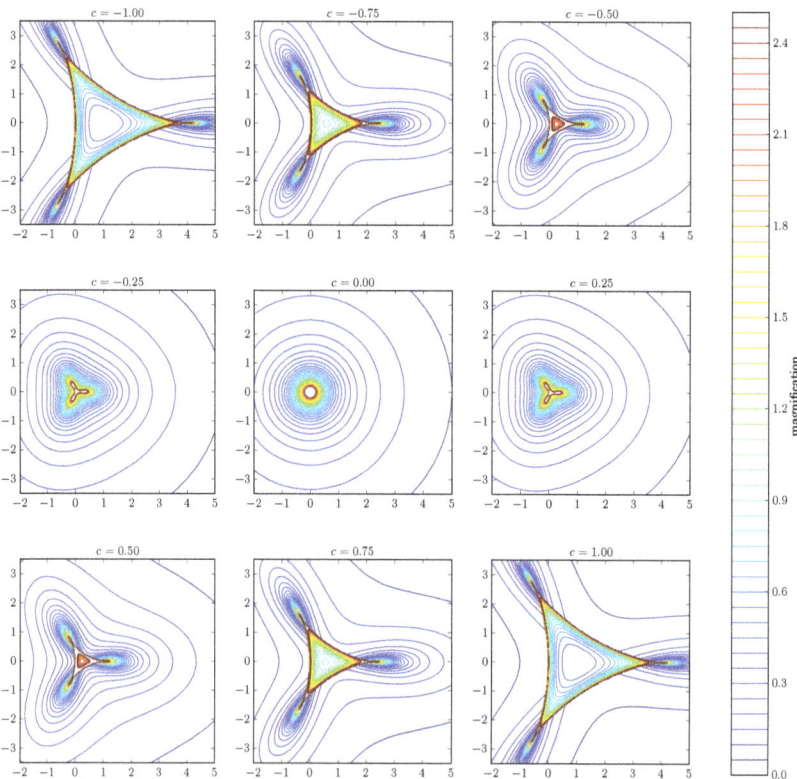

Figure 3. Level curves of the total unsigned magnification, for different values of c.

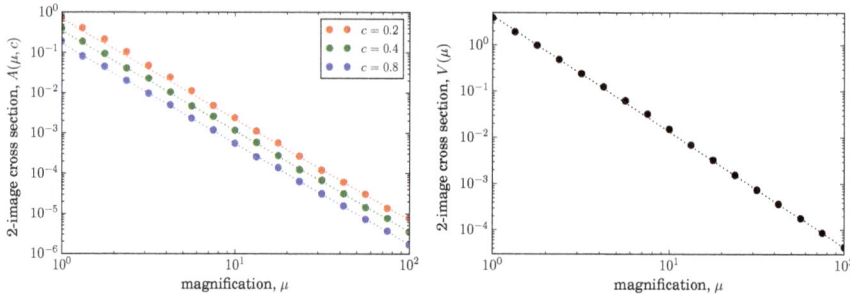

Figure 4. (**left**) The colored points show the area cross section $A(\mu, c)$ in the two-image region, computed numerically for different c-slices. The dotted lines show the scaling $A(\mu, c) \propto \mu^{-2.5}$. (**right**) The points show the volume cross section $V(\mu)$ in the two-image region, computed by integrating c over the range $[0.1, 1.0]$. The dotted line shows the scaling $V(\mu) \propto \mu^{-2.5}$.

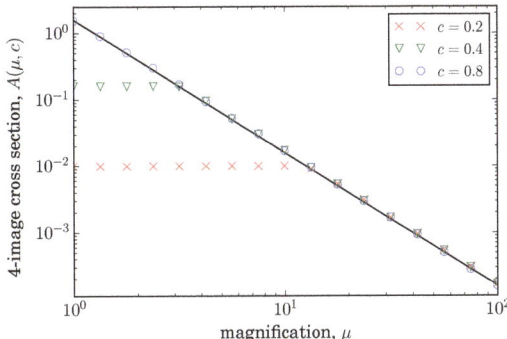

Figure 5. The colored points show the area cross section $A(\mu,c)$ in the four-image region, computed numerically for different c-slices. The solid line shows $A(\mu,c) = \pi/(2\mu^2)$, in agreement with the result obtained analytically in Equation (4) (with $\tilde{\mu} = \mu/2$). The cross section curves level off at the magnification that corresponds to the center of the caustic.

Observe that this area is c-independent. This area is related to $A(\mu,c)$, the magnification (area) cross section, as follows. Let $\vec{s}$ be a source in the four-image region whose total unsigned magnification is μ. Clearly $\vec{s} \notin S_\mu$, since its one positive-magnification image must be less than μ. However, it is known that the four-image region of the elliptic umbilic, for any $c \neq 0$, satisfies the following magnification relation.

$$\sum_{i=1}^{4} \tilde{\mu}_i = 0,$$

where $\tilde{\mu}_i$ is the signed magnification of lensed image i; see [19] for a proof. (For clarity, we use the notation "$\tilde{\mu}$" to denote the magnification belonging to an individual image, and reserve the notation "μ" to denote the total unsigned magnification of a source). It follows that $\vec{s} \in S_{\mu/2}$, because the positive-magnification image must have magnification $\mu/2$, since the other three magnifications are negative and must cancel it out. Thus, the closed curve $S_{\mu/2}$ is precisely the level curve of sources with total unsigned magnification μ. Observe that when $\tilde{\mu} = \mu/2$ in Equation (4), then $A(\mu,c) = \pi/(2\mu^2)$, as in Figure 5. And thus the area cross section scales like Equation (4). Figure 5 confirms this result numerically. Integrating Equation (4) from c_1 to c_2 yields a volume cross section that clearly scales as $V(\mu) \propto \mu^{-2}$.

There remains, finally, the two-image region outside the caustic curve shown in Figure 2. Here, for $\mu > 0$ large enough, those sources with total unsigned magnification greater than μ comprise the regions labeled A, B, and C, which three enclosed areas are equal by symmetry. In this case the area enclosed by them is not as easily derivable analytically, due to the complicated nature of the intersection points of the dashed curves with the caustic. However, it is possible to bound the areas A, B, and C, from above and below, and to show that these lower and upper bounds, which are functions of μ and c, both scale to leading order in μ as $\sim \mu^{-2.5}$, thereby supporting our numerical results in Figure 4. Figure 6 briefly describes the procedure, foregoing technical details.

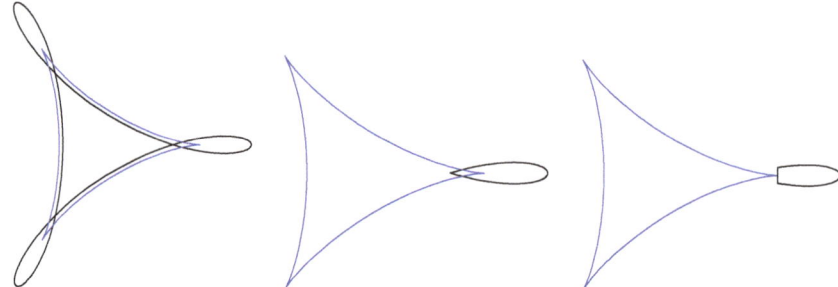

Figure 6. In each panel, the blue triangular-shaped curve is (a c-slice of) the elliptic umbilic caustic. In the **leftmost** figure, the black curve comprises all source positions with at least one image with (unsigned) magnification $\tilde{\mu}$, for some fixed value of $\tilde{\mu}$. In the large magnification limit, the magnification of one of the images in the two-image region dominates that of the other, in which case the loops enclosing each cusp in the leftmost panel approach the regions A, B, and C in Figure 2. The closed black curves in the **middle** and **rightmost** figures bound the areas A, B, and C from above and below. It can be shown that these two area bounds, which are functions of μ and c in general, both scale to leading order in μ as $\sim \mu^{-2.5}$, where μ is the total unsigned magnification of a source in the two-image region.

3. Conclusions

We have shown that the asymptotic scaling of the magnification volume cross section corresponding to an elliptic umbilic caustic surface is $\mu^{-2.5}$ in the two-image region and μ^{-2} in the four-image region, where μ is the total unsigned magnification. In both cases our results are supported both numerically and analytically. The goal was to extend to higher-order caustic singularities the well known (area) cross sections for fold (μ^{-2}) and cusp ($\mu^{-2.5}$) caustics, in particular given that the elliptic umbilic caustic can be manifested in both binary and three-galaxy lensing systems.

Author Contributions: A.B.A. and A.O.P. led the mathematical analysis; C.R.K. led the computational analysis.

Funding: This material is based upon work supported in part by the National Science Foundation under Grant No. 1641020.

Acknowledgments: Some of this work was done at the American Mathematical Society's June, 2018, Mathematics Research Conference on the "Mathematics of Gravity and Light".

Conflicts of Interest: The authors declare no conflict of interest.

References

1. Schneider, P.; Ehlers, J.; Falco, E. *Gravitational Lenses*; Springer: Berlin, Germany, 1992.
2. Kaiser, N. *New Insights into the Universe*; Martinez, V.J., Portilla, M., Saez, S., Eds.; Springer: Berlin, Germany, 1992.
3. Bartelmann, M.; Huss, A.; Colberg, J.M.; Jenkins, A.; Pearce, F.R. Arc statistics with realistic cluster potentials IV. Clusters in different cosmologies. *Astron. Astrophys.* **1998**, *330*, 1–9.
4. Petters, A.O.; Levine, H.; Wambsganss, J. *Singularity Theory and Gravitational Lensing*; Birkhäuser: Basel, Switzerland, 2001.
5. Lotz, J.M.; Koekemoer, A.; Coe, D.; Grogin, N.; Capak, P.; Mack, J.; Anderson, J.; Avila, R.; Barker, E.A.; Borncamp, D. The Frontier Fields: Survey design and initial results. *Astrophys. J.* **2017**, *837*, 97. [CrossRef]
6. Blandford, R.D.; Narayan, R. Fermat's principle, caustics, and the classification of gravitational lens images. *Astrophys. J.* **1986**, *310*, 568–582. [CrossRef]
7. Mao, S. Gravitational microlensing by a single star plus external shear. *Astrophys. J.* **1992**, *389*, 63–67. [CrossRef]
8. Schneider, P.; Weiss, A. The gravitational lens equation near cusps. *Astron. Astrophys.* **1992**, *260*, 1.

9. Arnold, V.I. Normal forms of functions near degenerate critical points, the Weyl groups of A_k, D_k, E_k and Lagrangian singularities. *Func. Anal. Appl.* **1973**, *6*, 3–25. [CrossRef]
10. Callahan, J. Singularities and plane maps. *Am. Math. Mon.* **1974**, *81*, 211–240. [CrossRef]
11. Callahan, J. Singularities and plane maps II: Sketching catastrophes. *Am. Math. Mon.* **1977**, *84*, 765–803. [CrossRef]
12. Majthay, A. *Foundations of Catastrophe Theory*; Pitman: Wetherby, UK, 1985.
13. Arnold, V.I.; Gusein-Zade, S.M.; Varchenko, A.N. *Singularities of Differentiable Maps, Volume 1*; Birkhäuser: Basel, Switzerland, 1985.
14. Petters, A.O. Arnold's singularity theory and gravitational lensing. *J. Math. Phys.* **1993**, *33*, 3555. [CrossRef]
15. Rusin, D.; Kochanek, C.S.; Norbury, M.; Falco, E.E.; Impey, C.D.; Lehár, J.; McLeod, B.A.; Rix, H.-W.; Keeton, C.R.; Muñoz, J.A.; et al. B1359+154: A Six-Image Lens Produced by a $z \simeq 1$ Compact Group of Galaxies *Astrophys. J.* **2001**, *557*, 594. [CrossRef]
16. Blandford, R.D. The Future of Gravitational Optics. *Publ. Astron. Soc. Pac.* **2001**, *113*, 789. [CrossRef]
17. Shin, E.M.; Evans, N.W. Lensing by binary galaxies modelled as isothermal spheres. *Mon. Not. R. Astron. Soc.* **2008**, *390*, 505–522. [CrossRef]
18. Orban de Xivry, G.; Marshall, P. An atlas of predicted exotic gravitational lenses. *Mon. Not. R. Astron. Soc.* **2009**, *399*, 2–20. [CrossRef]
19. Aazami, A.B.; Petters, A.O. A universal magnification theorem for higher-order caustic singularities. *J. Math. Phys.* **2009**, *50*, 032501. [CrossRef]

© 2019 by the authors. Licensee MDPI, Basel, Switzerland. This article is an open access article distributed under the terms and conditions of the Creative Commons Attribution (CC BY) license (http://creativecommons.org/licenses/by/4.0/).

Article

Caustics in Gravitational Lensing by Mixed Binary Systems

Valerio Bozza [1,2,*], Silvia Pietroni [1,2] and Chiara Melchiorre [3]

[1] Dipartimento di Fisica "E.R. Caianiello", Università di Salerno, Via Giovanni Paolo II 132, I-84084 Fisciano, Italy; spietroni@unisa.it
[2] Istituto Nazionale di Fisica Nucleare, Sezione di Napoli, Via Cintia, 80126 Napoli, Italy
[3] Miur: Ministero dell'Istruzione, dell'Università e della Ricerca, 00153 Roma, Italy; chiara.melchiorre@istruzione.it
* Correspondence: valboz@sa.infn.it

Received: 22 June 2020; Accepted: 29 July 2020; Published: 31 July 2020

Abstract: We investigated binary lenses with $1/r^n$ potentials in the asymmetric case with two lenses with different indexes n and m. These kinds of potentials have been widely used in several contexts, ranging from galaxies with halos described by different power laws to lensing by wormholes or exotic matter. In this paper, we present a complete atlas of critical curves and caustics for mixed binaries, starting from the equal-strength case, and then exploring unequal-strength systems. We also calculate the transitions between all different topology regimes. Finally we find some useful analytic approximations for the wide binary case and for the extreme unequal-strength case.

Keywords: gravitational lensing; black holes; wormholes; galaxies

1. Introduction

Space–time is curved by the presence of massive bodies and this curvature influences the motion of the bodies themselves: this leads to a geometry in constant evolution. One of the consequences is that even light, supposed to be massless, bends its trajectory while passing close to a massive body. Einstein deduced it already in 1913, two years before his theory was completed [1], and the British astronomer Arthur Eddington decided to exploit this intuition experimentally. On 29 May 1919 during a solar eclipse in Principe Islands he showed that stars moved from their position by the amount precisely predicted by general relativity. This great result was put into evidence by the main newspapers of that time, like *Cosmic Time* that titled "Sun's gravity bends starlight" underlining the triumph of Einstein's theory. This was the first observation of *gravitational lensing* [2].

Gravitational lensing is an important tool in astrophysics and in cosmology widely used to study both populations of compact objects (including exoplanets, black holes, and other stellar remnants) [3,4], and extended objects, such as galaxies, clusters of galaxies, and large-scale structures [5–9]. Since most of the mysteries of our Universe do not show up in observations based on electromagnetic interactions, gravitational lensing is more and more employed to study the dark side of the Universe, including dark matter, dark energy, and any kind of exotic matter (such as wormholes) conjectured by theorists [10–15].

Gravitational lensing effects by wormholes were investigated in [16,17], with negative mass in [18–22], and with positive mass in [23–27]. We want to remark that in 1973, Ellis and Bronnikov independently found a massless wormhole (the Ellis wormhole) as a wormhole solution of the Einstein equations, see [28,29]. Spherically symmetric and static traversable Morris–Thorne wormholes were analyzed in [30,31]. The most general extension of the Morris–Thorne wormhole is the solution of the stationary and axially symmetric rotating Teo wormhole in [32], the first rotating wormhole solution, and this was the starting point for the investigation of gravitational lensing by rotating wormholes explored by Jusufi and Ovgun in [33]. Tsukamoto and Harada studied the light rays passing through a wormhole in [34]; Ohgami and Sakai studied the images of wormholes surrounded by optically thin dust in [35] in order to state if it is possible to identify wormholes by observing shadows; this was also investigated in [36] in rotating dust flow.

The metric of the Ellis wormhole falls down asymptotically as $1/r^2$ and its deflection angle goes as the inverse square of the impact parameter $1/u^2$ as explored in [37–45]. Metrics falling as $1/r^n$ were investigated also by Kitamura et al. [46] who found out that the deflection angle falls down with the same exponent as the metric: $\hat{\alpha} \sim 1/u^n$ with $n > 1$. Other investigations include [47–52]. Power-law deflection terms can also be found in gravitational lensing in the presence of plasma [53–57]. Particular attention was posed on the study of caustics of $1/r^n$ binary lenses by Bozza and Melchiorre in [58] and to the investigation of gravitational lensing by exotic lenses with a non-standard form of the equation of state or with a modified gravity theory by Asada [59]. A new method of detecting Ellis wormholes by the use of the images of wormholes surrounded by optically thin dust was investigated by Ohgami and Sakai [35].

After the Event Horizon Telescope results [60], consisting in the detection of the shadow of a supermassive black hole in the center of galaxy M87, many authors tried to explore new frontiers, and an interesting new reference is from Tsukamoto and Kokubu [61]: they investigate the collision of two test particles in the Damour–Solodukhin wormhole spacetime where Damour and Solodukhin stated in [62] that is not possible to distinguish black holes from wormholes with observations on a limited timescale.

From the side of binary galaxies as binary lenses we must cite the considerable work of Shin and Evans [63] that discussed the critical curves and caustics in the case $n < 1$. This applies to generic galactic halos and isothermal sphere in particular, as the limit $n \to 0$.

Kovner investigated extremal solutions for a singular isothermal sphere with a tide (SIST) [64]; Evans and Wilkinson studied lens models for representing cusped galaxies and clusters, as isothermal cusps always generate a pseudocaustic [65], while Rhie discussed pseudocaustics of various lens equations [66]. Wang and Turner studied strong gravitational lensing by spiral galaxies, modeling them as infinitely thin uniform disks embedded in singular isothermal spheres [67], while Tessore and Metcalf investigated a general class of lenses following an elliptical power law profile [68]. All these systems possess pseudocaustics that were also investigated by Lake and Zheng in gravitational lensing by a ring-like structure [69]. Higher-order caustic singularities, such as the elliptic umbilic, were discussed by Aazami et al. [70].

In this work we want to extend the symmetric structure already studied by Bozza and Melchiorre in [58] for $1/r^n$ potentials, in which the two lenses have the same index n, to an asymmetric case in which the lenses have different indexes. This generalization is particularly useful in both scientific contexts described by $1/r^n$ potentials. In fact, we may have pairs of galaxies that have very different structures and thus different halo profiles, e.g., a dwarf galaxy as a satellite to a giant galaxy. On the other hand, if wormholes or other exotic objects exist, they might be part of a binary system with an ordinary star or other compact objects. The co-existence of objects with different $1/r^n$ potentials thus seems plausible in many situations, thus justifying the generalization we are going to undertake here.

In Section 2 we give the lens equation for $1/r^n$ potentials for two exotic lenses with different n. In Section 3 we study critical curves and caustics presenting three main cases: equal-strength binary lenses, unequal-strength binary and extreme unequal-strength binary lenses explaining the origin of the pseudocaustic and of the elliptic umbilic catastrophe for $mn < 1$. In Section 4 we study the transitions between different caustic topologies. In Section 5 we derive analytical approximations for the three cases analyzed in Section 3 in order to have a deeper understanding in the caustic evolution, in its shape and size. Finally in Section 6 we draw our conclusions.

2. Gravitational Lensing by Objects with $1/r^n$ Potential

Objects whose gravitational potential asymptotically falls as $1/r^n$ ($n \leq 1$ for ordinary matter, $n > 1$ for exotic matter) give rise to a deflection angle that goes as $\alpha \sim 1/|\theta|^n$, where θ is the angular position at which the image is observed. The lens equation for a single lens, first studied by Kitamura et al. in [46] and then generalized by Bozza and Postiglione in [52], is

$$\beta = \theta - \frac{\theta_E^{n+1}}{|\theta|^n} Sign(\theta),\tag{1}$$

where β is the source angular position with respect to the center of the lens, θ_E is the Einstein radius of the lens, which depends on the specific parameters of the metric describing the object [46] and the index n is either the exponent of the halo profile for a normal matter distribution or the ratio between tangential and radial pressure, $n = -2p_t/p_r$, if we consider exotic matter [52].

We want to explore a system composed by two objects in the asymmetric case in which our lenses have different indexes, here indicated with n and m. The binary lens equation is

$$\vec{\beta} = \vec{\theta} - \theta_{E,A}^{n+1} \frac{\vec{\theta} - \vec{\theta}_A}{|\vec{\theta} - \vec{\theta}_A|^{n+1}} - \theta_{E,B}^{m+1} \frac{\vec{\theta} - \vec{\theta}_B}{|\vec{\theta} - \vec{\theta}_B|^{m+1}},\tag{2}$$

where $\vec{\theta}_A$ and $\vec{\theta}_B$ are the coordinates of the two objects in the sky.

We note that the Einstein radii $\theta_{E,A}$ and $\theta_{E,B}$ appear with different exponents for each lens. It is thus convenient to use $\theta_{E,A}$ as a unit of measure for angles and define the "strength ratio" as $\gamma = \theta_{E,B}/\theta_{E,A}$. We rewrite the lens equation as follows

$$\vec{\beta} = \vec{\theta} - \frac{\vec{\theta} - \vec{\theta}_A}{|\vec{\theta} - \vec{\theta}_A|^{n+1}} - \gamma^{m+1} \frac{\vec{\theta} - \vec{\theta}_B}{|\vec{\theta} - \vec{\theta}_B|^{m+1}}.\tag{3}$$

Now we introduce complex coordinates [71]

$$\zeta = \beta_1 + i\beta_2; z = \theta_1 + i\theta_2 \tag{4}$$

We take the mid-point between the two lenses as the origin of the coordinates, and orient the real axis along the line joining the two lenses. We thus set $z_A = -s/2$ and $z_B = s/2$, where s is the normalized angular separation between the lenses. The lens equation becomes

$$\zeta = z - \frac{1}{\left(z+\frac{s}{2}\right)^{\frac{n-1}{2}} \left(\bar{z}+\frac{s}{2}\right)^{\frac{n+1}{2}}} - \frac{\gamma^{m+1}}{\left(z-\frac{s}{2}\right)^{\frac{m-1}{2}} \left(\bar{z}-\frac{s}{2}\right)^{\frac{m+1}{2}}} \tag{5}$$

The Jacobian determinant of the lens map in complex notation is given by

$$J(z,\bar{z}) = \left|\frac{\partial \zeta}{\partial z}\right|^2 - \left|\frac{\partial \zeta}{\partial \bar{z}}\right|^2,\tag{6}$$

which in our case becomes

$$J = \left[1 + \frac{1}{2}\left(\frac{n-1}{\left(z+\frac{s}{2}\right)^{\frac{n+1}{2}} \left(\bar{z}+\frac{s}{2}\right)^{\frac{n+1}{2}}} + \frac{\gamma^{m+1}(m-1)}{\left(z-\frac{s}{2}\right)^{\frac{m+1}{2}} \left(\bar{z}-\frac{s}{2}\right)^{\frac{m+1}{2}}}\right)\right]^2 \\ - \frac{1}{4}\left|\frac{n+1}{\left(z+\frac{s}{2}\right)^{\frac{n+3}{2}} \left(\bar{z}+\frac{s}{2}\right)^{\frac{n-1}{2}}} + \frac{\gamma^{m+1}(m+1)}{\left(z-\frac{s}{2}\right)^{\frac{m+3}{2}} \left(\bar{z}-\frac{s}{2}\right)^{\frac{m-1}{2}}}\right|^2. \tag{7}$$

We note that the structure of the Jacobian becomes more complicated with respect to the ordinary point-lenses ($m = n = 1$), in which many terms disappear. We thus expect a correspondingly richer phenomenology. The Schwarzschild case was already explored by Schneider and Weiss in [72] for lenses with the same mass and by Erdl and Schneider for lenses with different masses [73]; Bozza and Melchiorre investigated the case $m = n$. In order to compare our results with theirs, it is important to note that there is no notion of a combined total Einstein radius when the two lenses have different indexes for their potentials. Therefore, the notation introduced there, with ϵ_i as the ratio of the individual lens strength to the total strength cannot be replicated here. Their results were expressed in terms of the ratio $q = \epsilon_B/\epsilon_A$. The relation between our parameter $\gamma = \theta_{E,B}/\theta_{E,A}$ and q is just $\gamma^{m+1} = q$. As a practical example, the Einstein radius scales as $\sqrt{q}$ in the Schwarzschild case, where q becomes the mass ratio of the two lenses.

3. Critical Curves and Caustics

The condition $J(z) = 0$ defines the critical curves on the lens plane. By applying the lens map on critical points we find the corresponding points on the source plane, which form the caustics. Critical curves and caustics are of fundamental importance to understand how gravitational lensing works. When a source crosses a caustic, a new pair of images is created on the corresponding point in the critical curve. Therefore, caustics bound regions with a different number of images. Critical curves distinguish regions in which images have opposite parities.

Our model contains four parameters: the indexes of the two potentials n, m, the separation between the two lenses s, and the ratio of the two Einstein radii γ. In order to start the exploration of this parameter space, we first analyze the equal-strength case with $\gamma = 1$, and then move to unequal strength cases.

In all plots presented in this paper, we keep $n = 1$ fixed for the first lens (ordinary Schwarzschild lens), with variable m for the second lens: $m = 0, 0.5, 1, 2, 3$ (we remind that $m = 0$ is the singular isothermal sphere, already investigated by Shin and Evans in [63], galactic halos are in the range $0 < m < 1$ and $m = 2$ corresponds to the Ellis wormhole; objects with $m > 1$ require exotic matter).

Critical curves are obtained by the contour plot of the Jacobian determinant and these contours are then mapped through the lens equation in order to get the caustics. All computations are performed by Wolfram Mathematica 11[1].

3.1. Equal-Strength Binaries

In the equal-strength case, we set $\gamma = 1$, which means that $\theta_{E,A} = \theta_{E,B}$: both lenses would generate a critical curve with the same radius if they were isolated.

For the standard binary Schwarzschild lens [72], we know that three topologies exist:

- close separation, for $s < s_{CI}$;
- intermediate separation, for $s_{CI} < s < s_{IW}$;
- wide separation, for $s > s_{IW}$;

and the two transitions are $s_{CI} = 1$ and $s_{IW} = 2\sqrt{2}$ in our units.

We find that these three topologies persist for any values of n and m, although the boundary values may vary somewhat. In order to illustrate the evolution of critical curves and caustics in intelligible figures, we present the plots for different values of m at fixed values of separation s, starting from wide separation binaries and then moving the two lens closer.

First, in Figure 1, we have two lenses at wide separations for $s = 3.4$. Here we clearly see how the Einstein ring of each lens is distorted by the presence of the partner lens. Comparing the critical curves obtained at different values of the index m, we clearly see that the distortion is stronger for small values of m. This is a direct consequence of the fact that the potential decays more steeply for larger m and thus the first lens feels a weaker tidal field from the second lens. This is particularly evident for the caustic of the first lens, which becomes very small at $m = 3$, while it becomes larger and more shifted at $m = 0$. The caustic of the second lens is almost independent of m. In practice,

---

[1] https://www.wolfram.com/mathematica/

the shape and the size of the caustic is mostly determined by the tidal field of the first lens, which we are keeping fixed with $n = 1$.

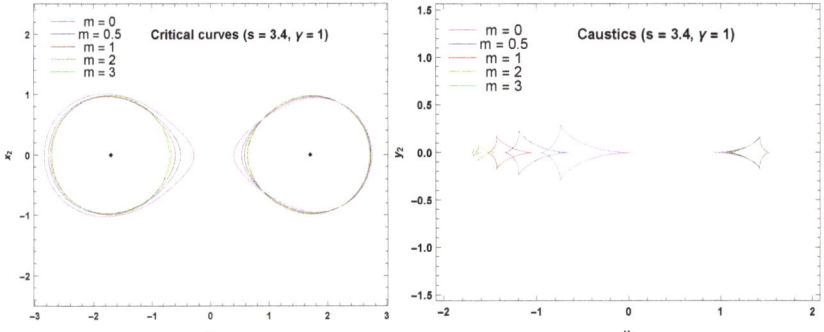

Figure 1. Critical curves and caustics in the equal-strength binary, wide separation. Here and in the following figures the lens on the left side has $n = 1$ and the lens on the right side has variable m, coherently with Equation (5).

In Figure 2 we show critical curves and caustics for $s = 2\sqrt{2}$, which corresponds to the intermediate-wide transition in the standard $n = m = 1$ case. In fact, the red curves show the typical beak-to-beak singularity in the origin. For $m < 1$ we are already in the intermediate regime, while for $m > 1$ we are still in the wide regime. As explained before, the fact that the intermediate regime extends to larger separations for $m < 1$ is a consequence of the slower decay of the potential.

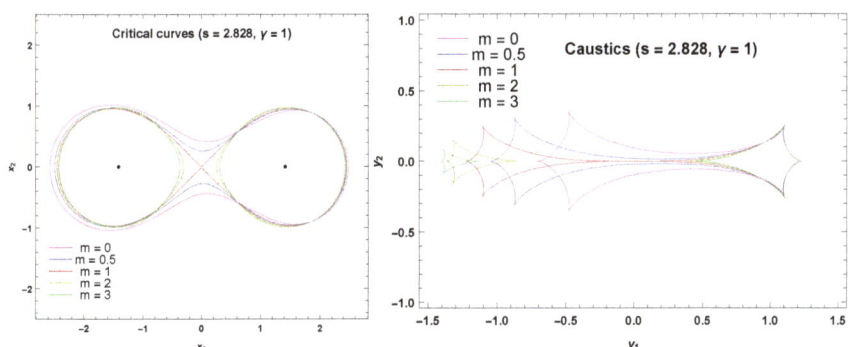

Figure 2. Critical curves and caustics in the equal-strength binary, intermediate-wide transition.

In Figure 3 we see an intermediate separation at $s = 1.4$: critical curves are larger for smaller values of m, while caustics are larger for increasing m. Indeed, we are starting to see some kind of inversion in the behavior of the lenses. Steeper profiles are going to dominate at smaller separations, as will be more evident in the incoming figures.

In Figure 4 we show the critical curves and caustics at $s = 1$, which corresponds to the close-intermediate transition for the standard $n = m = 1$ case. In fact, the red curve shows the two symmetric beak-to-beak singularities. Contrary to the previous transition, now the $m < 1$ caustics are already in the close regime, with small oval critical curves generating small triangular caustics. The $m > 1$ curves are still in the intermediate regime. Following the same reasoning, $m < 1$ lenses become subdominant in this regime and their influence on the whole system is smaller. In this regime, we also find the *elliptic umbilic catastrophe* that we shall discuss in Section 3.1.2.

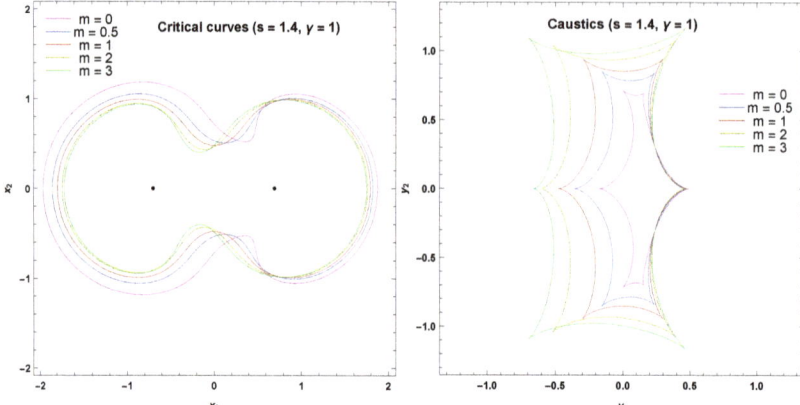

Figure 3. Critical curves and caustics in the equal-strength binary, intermediate separation.

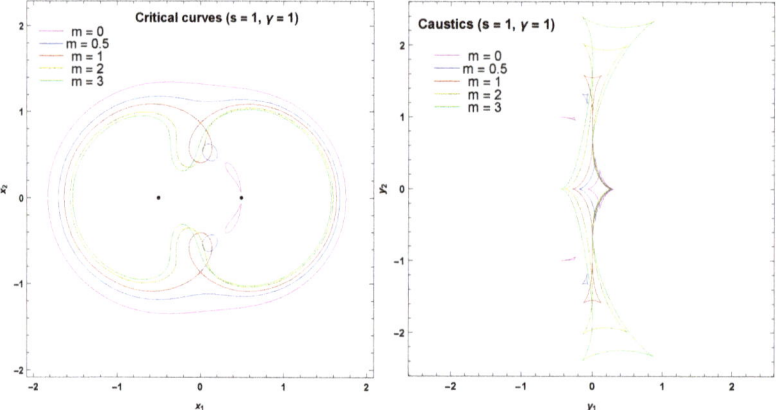

Figure 4. Critical curves and caustics in the equal-strength binary, close-intermediate transition.

In Figure 5 we can see critical curves and caustics in the close separation, for $s = 0.8$.

Primary critical curves are big ovals that become smaller as m increases. In fact, $m > 1$ curves tend to be closer to an intermediate regime. Secondary critical curves are small ovals that move far from the second lens, in the left direction, for $m > 1$; for $m = 0$ (magenta line) they converge on the second lens with the shape of a lemniscate: in this point the lens map is indeterminate and the corresponding caustics remain open on a circle that is called pseudocaustic. We shall discuss this structure in Section 3.1.1.

On the other side we find that the central caustics have the typical 4-cusps shape and they become smaller as m decreases. Secondary caustics are always triangular but are considerably larger for $m > 1$, a fact that was already stressed in [58]. Note that for $m > 1$ triangular caustics move right, while the central caustic is slightly displaced to the left. The opposite occurs for $m < 1$. We can find a similar behavior for standard Schwarzschild binaries with unequal masses. In practice, although we started with the same Einstein radius for both lenses, steeper profiles ($m > 1$) behave similarly to heavier masses in this regime, while shallower profiles ($m < 1$) behave as lighter masses.

3.1.1. The Pseudocaustic

A pseudocaustic is a closed curve on the source plane that exists for singular distributions with zero core radius. In the singular limit, the radial critical curve collapses onto the center of the lens, leaving no space for the dim central type III image. When the source crosses the corresponding radial caustic, only one more image forms, while the other image is degenerate with the center of the lens. The radial caustic is then named pseudocaustic, since it behaves differently from normal caustics [64–69].

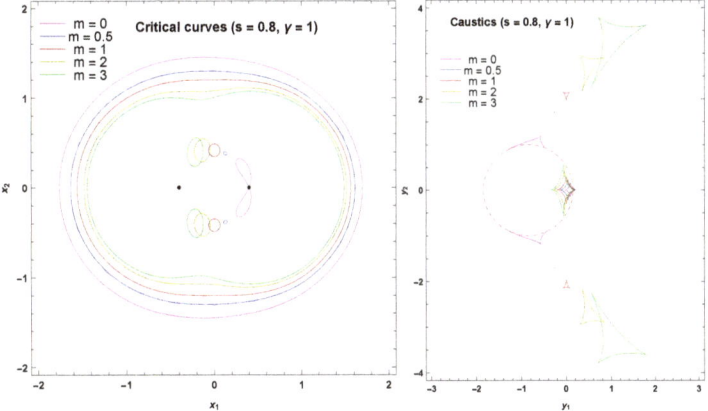

Figure 5. Critical curves and caustics in the equal-strength binary, close separation. Dashed magenta circle indicates the pseudocaustic for $m = 0$.

In the binary case, a pseudocaustic may still exist in the singular limit $m = 0$. Through an analytical exploration we find out the points where the two secondary triangular caustics touch the pseudocaustic.

The pseudocaustic is generated by critical curves collapsing to the center of the lens when $m = 0$. In order to explore what happens around the center of the second lens, we set

$$z = \frac{s}{2} + \epsilon_1 + i\epsilon_2; \qquad (8)$$

we expand around zero at $1/\epsilon$ order and then we solve with respect to ϵ_2.

We get two symmetric solutions

$$\epsilon_2 = \pm \frac{\sqrt{s^2+1}}{\sqrt{1-s^2}} \epsilon_1 \qquad (9)$$

These solutions are two straight lines that cross at the origin of the system, and their angular coefficient is real only for $s < 1$. This means that the two small oval critical curves will touch the center of the $m = 0$ lens for separations in this regime. By substituting in the lens equation we find

$$\zeta = \frac{s}{2} - \frac{1}{s} \pm \gamma \sqrt{-s^2 \pm \sqrt{s^4-1}} \qquad (10)$$

These are the coordinates of the four contact points of the two triangular caustics with the pseudocaustic of radius γ and center $(\frac{s}{2} - \frac{1}{s}, 0)$. The term $\frac{1}{s}$ shifts the caustic to the left side with respect to the position of the second lens $\frac{s}{2}$.

If the source only crosses the pseudocaustic, we have the sudden creation of one image of negative parity; if the source crosses a triangular caustic first and then the pseudocaustic, we have the formation of two images and the one inside the lemniscate (with positive parity) collapses on the lens.

3.1.2. The Elliptic Umbilic

As shown in [58,63], in the range $0 \leq m < 1$ an *elliptic umbilic catastrophe* exists in the close separation. In an elliptic umbilic, the size of the small oval critical curves goes to zero and then grows up to finite size again. The catastrophe lies on a circle centered in the origin of the system, at the mid-point between the two lenses, and passing through them. It occurs at a specific separation s, which depends on the other parameters of the lens γ, m, n.

To find out the separation s, for any m and n, at which the catastrophe occurs we proceed as follow: first we write the system of equations

$$\begin{cases} J = 0 \\ \frac{\partial J}{\partial z} = 0 \end{cases} \tag{11}$$

along the circle, i.e., we set

$$z = s \frac{e^{i\theta}}{2}. \tag{12}$$

Then we introduce a new angular variable t in order to simplify our computation

$$t = \frac{\sin^{m+1}(\theta/2)}{\cos^{n+1}(\theta/2)} \tag{13}$$

From Equation (11), we get the angular position of the elliptic umbilic

$$t = \frac{(m+1)\gamma^{m+1}}{(n+1)s^{n-m}} \tag{14}$$

and then we finally obtain the value of s at which the catastrophe happens

$$s_{euc} = \left(\frac{1-mn}{m+1}\right)^{\frac{1}{n+1}} \sqrt{1 + \frac{\gamma^2(m+1)^{\frac{2}{n+1}}}{(n+1)^{\frac{2}{m+1}}(1-mn)^{\frac{2(m-n)}{(m+1)(n+1)}}}} \tag{15}$$

Note that the solution exists for $mn < 1$. In mixed binaries, we may have an elliptic umbilic also when one of the two lenses has a steep potential with $n > 1$. In order to illustrate this, we choose $n = 2$ (exotic matter) and $m = 0.25$ (a possible galactic halo). We can see, in Figure 6, a zoom on the small oval critical curve for $0.714 \leq s \leq 0.834$ in steps of 0.02. The separation at which the catastrophe occurs is $s_{euc} = 0.774$. The critical curve shrinks to zero size for growing s, from $s = 0.714$ to $s = 0.774$ (lower curves) and then it grows up again. The corresponding triangular caustics behave similarly.

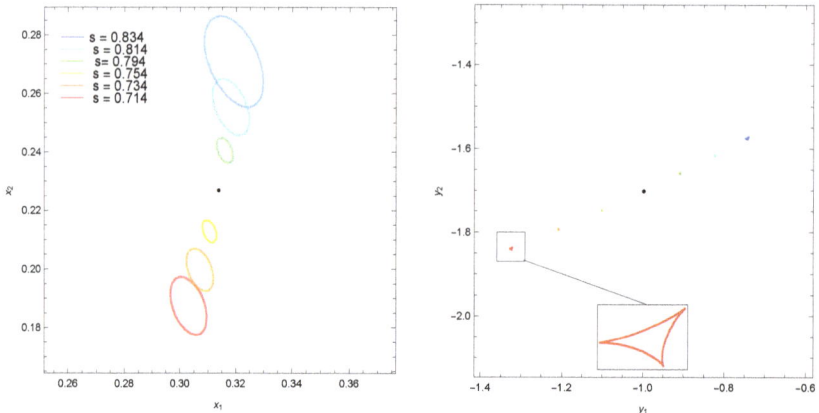

Figure 6. The elliptic umbilic catastrophe for $n = 2$, $m = 0.25$ and $0.714 \leq s \leq 0.834$ in steps of 0.02, one color for each s from red to blue. The separation at which the catastrophe occurs is $s_{euc} = 0.774$. Critical curves on the left side panel, caustics on the right side panel.

3.2. Unequal-Strength Binary

In the unequal-strength binary case, in order to keep contact with the previous work, we consider $q = 0.1$, as in [58], so in terms of the ratio of the Einstein radii, our strength ratio is $\gamma = \sqrt{0.1}$. We need to multiply s in [58] by a factor $\sqrt{q+1} = \sqrt{1.1}$, so the transitions between different topologies for $n = m = 1$ occur as follows:

- close-intermediate transition, $s_{CI} = 0.807$;
- intermediate-wide transition, $s_{IW} = 1.772$.

Therefore, in this subsection we have the standard lens on the left with bigger Einstein radius than the lens on the right, for which we vary the potential index m.

We shall discuss each value of m in the range $0 \leq m \leq 3$ in detail.

In Figure 7 we show the wide separation for $s = 2.1$. Critical curves are separated and slightly deformed. The caustic of the left side lens is smaller since the tidal field from the right side lens is normally weaker. However, for $m < 1$ the potential decays slower enough to make the left side caustic bigger than the caustic on the right.

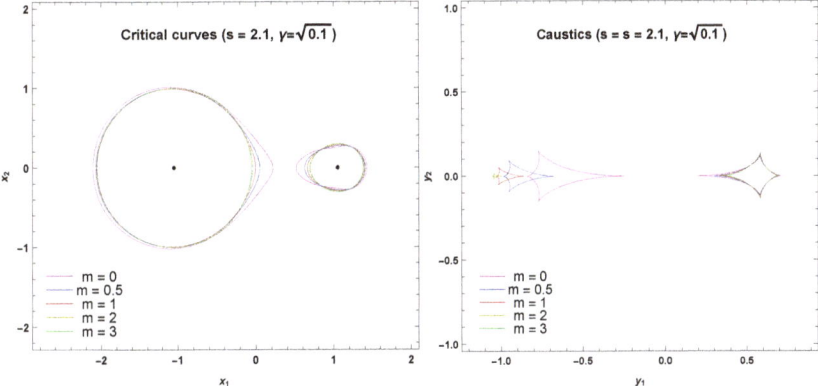

Figure 7. Critical curves and caustics in the unequal-strength binary, wide separation. Here and in the following figures the lens on the left has $n = 1$ and the lens on the right has variable m.

In Figure 8 we can see the intermediate-wide transition at $s = 1.772$. For $m < 1$ (magenta and blue lines) the transition to the intermediate regime has already occurred, while we are still in the wide regime for $m > 1$ (green and yellow lines).

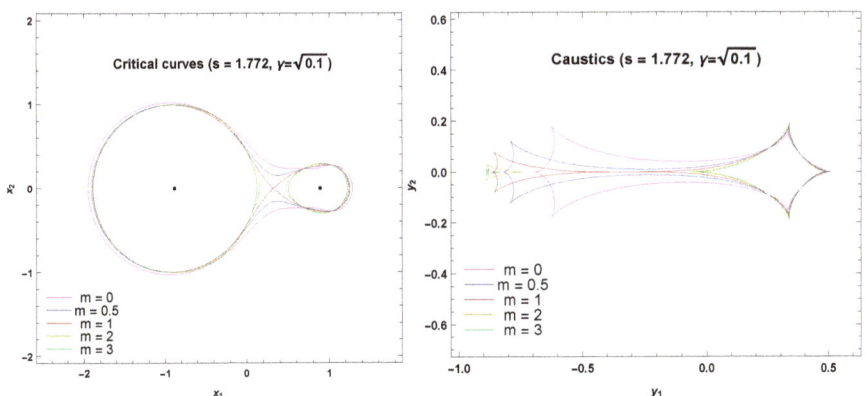

Figure 8. Critical curves and caustics in the unequal-strength binary, intermediate-wide transition.

In Figure 9 we can see the intermediate separation for $s = 1.05$. Critical curves now are all joined and they get smaller with increasing m. Caustics now have the 6-cusps shape and they get smaller with decreasing m. Note that the throat of the critical curve is wider for $m > 2$ and narrower for $m < 1$. Correspondingly, the fold between the off-axis cusps is longer for $m > 1$ and is extremely short for $m = 0$, where the two off-axis cusps almost coincide.

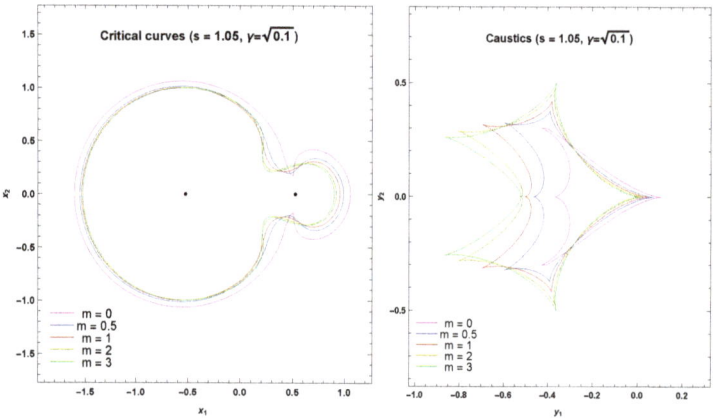

Figure 9. Critical curves and caustics in the unequal-strength binary, intermediate separation.

In Figure 10 we have the close-intermediate transition for $s = 0.807$. For $m = 0$ and $m = 0.5$, the primary caustics are already in the close regime, with the smaller ovals detached from the primary critical curve; for $m = 1$ we see the transition (red line), for $m > 1$ we are still in the intermediate regime. Note that the $m = 0$ ovals already reached the right side lens and the triangular caustics reached the pseudocaustic.

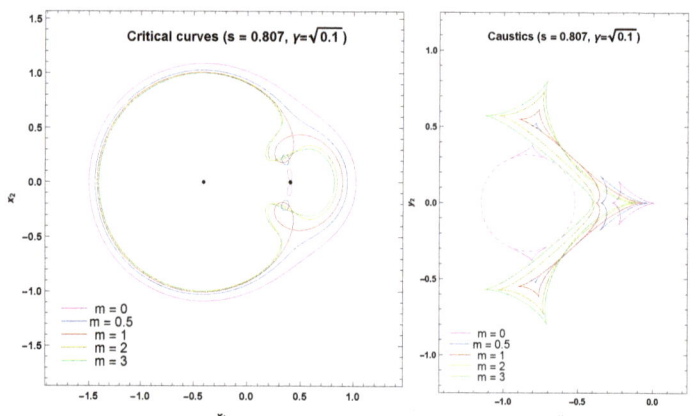

Figure 10. Critical curves and caustics in the unequal-strength binary, close-intermediate transition. Dashed magenta circle indicates the pseudocaustic for $m = 0$.

In Figure 11 we show the close separation for $s = 0.63$: the main critical curves, that generate the central caustics, are big ovals growing up in size with decreasing m. Secondary critical curves are small ovals close to the second lens, moving in the left direction as m increases. Like the equal-strength ratio case, for $m = 0$ the secondary critical curves are attached in a lemniscate shape and the corresponding caustics remain open on the pseudocaustic (see Section 3.1.1).

On the right panel we have the caustics: as m decreases, the central caustic moves to the right; secondary caustics become larger for greater values of m.

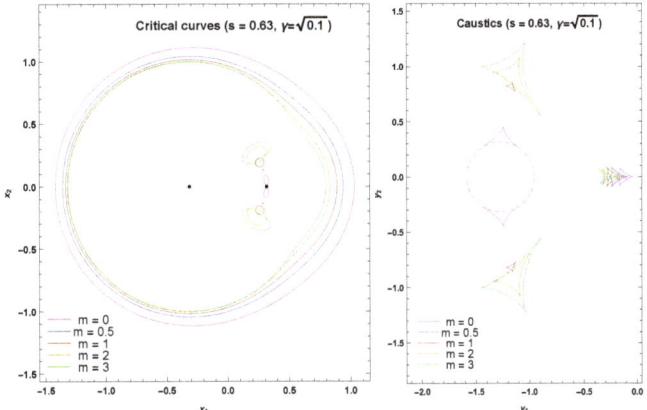

Figure 11. Critical curves and caustics in the unequal-strength binary, close separation. The dashed magenta circle indicates the pseudocaustic for $m = 0$.

3.3. Reversed Unequal-Strength Binary

In the previous section we assumed that the bigger lens was standard ($n = 1$) and the smaller lens had a different index m. In this section we study the reverse situation: the standard lens is smaller and the other lens is bigger. We thus keep $\gamma = \sqrt{0.1}$, fix $m = 1$ and let n vary.

In Figure 12 we start from the wide separation. Similarly to Figure 7, the caustic of the non-standard lens remains unaffected, while the caustic of the standard object strongly depends on the tidal field of the other lens. The shift and the size are much more affected than before, since now the standard lens is the weaker one.

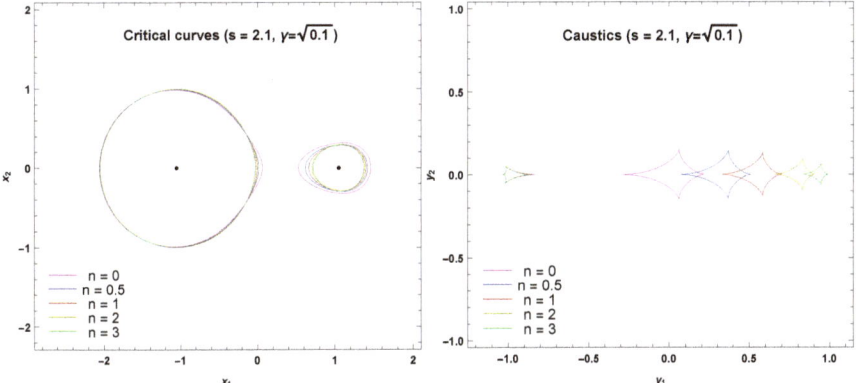

Figure 12. Critical curves and caustics in the unequal-strength binary with the standard lens on the right, wide separation. Here and in the following figures the lens on the right has $m = 1$ and the lens on the left has variable n.

In Figure 13, we are at the intermediate-wide transition. The situation is quite similar to Figure 8, with stronger dependence on the index n, as discussed before.

Figure 14 shows the intermediate topology. Comparing with Figure 9, it is interesting to note that here the left cusp is common for all caustics, while there it was the right cusp to be shared among all caustics. Of course, we can still interpret this fact through the variations of the tidal fields.

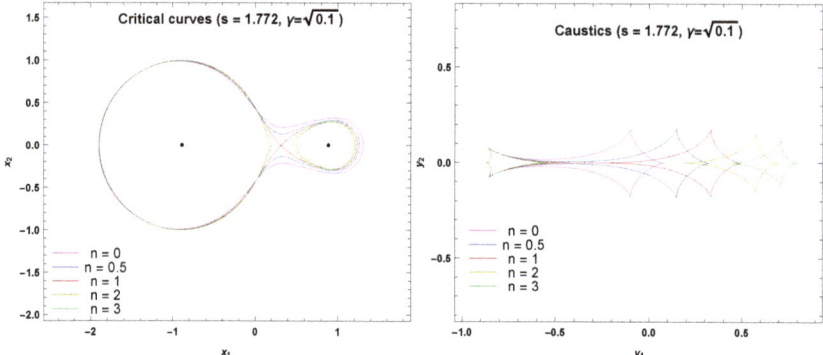

Figure 13. Critical curves and caustics in the unequal-strength binary with the standard lens on the right, intermediate-wide transition.

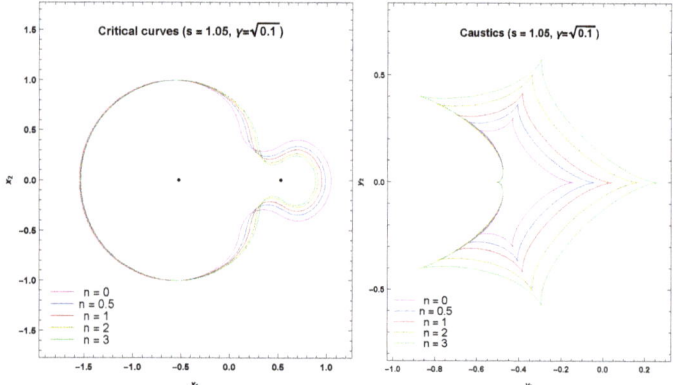

Figure 14. Critical curves and caustics in the unequal-strength binary with the standard lens on the right, intermediate separation.

Figure 15 shows the close-intermediate transition. Note that the red curves ($n = m = 1$) are exactly at the transition, while both larger and smaller n curves are in the close regime. This is not what happens in Figure 10, where larger m curves were still in the intermediate regime. Then we learn that the close regime is more extended for all $n \neq 1$ in this case.

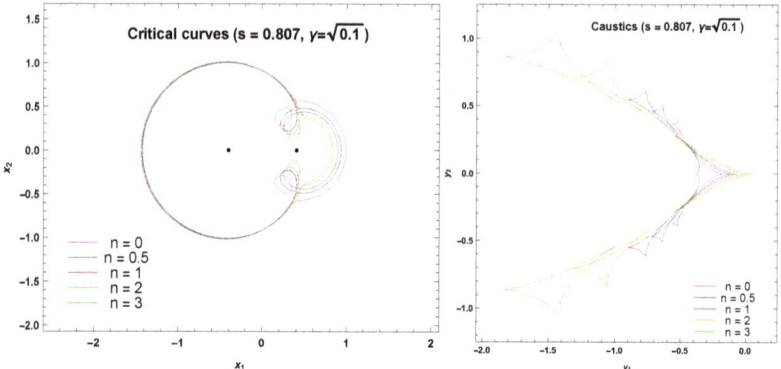

Figure 15. Critical curves and caustics in the unequal-strength binary with the standard lens on the right, close-intermediate transition.

Finally, Figure 16 shows the close regime. Note that the $n = 0$ small ovals do not collapse to the left lens but remain quite far. The pseudocaustic is never reached by the triangular caustics.

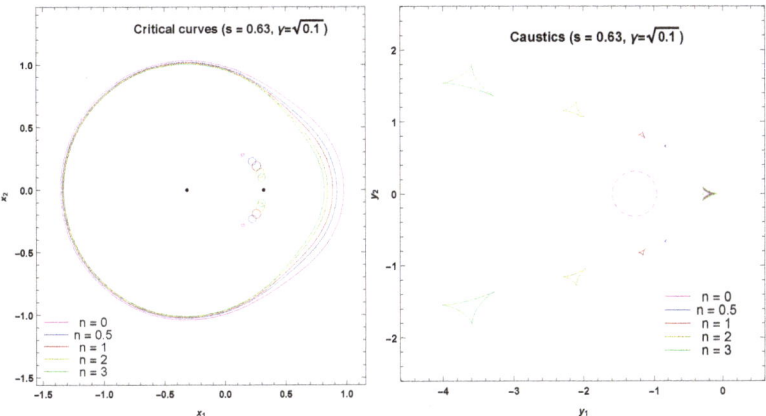

Figure 16. Critical curves and caustics in the unequal-strength binary with the standard lens on the right, close separation.

4. Transitions between Different Topologies

Now we want to find out the boundaries for the three topology regimes, s_{CI} and s_{IW}, for any n, m and γ.

As we know that transitions occur via higher order singularities of the lens map (beak-to-beak singularity in the binary lens case), in order to find out s_{CI} and s_{IW} we need to solve again the system of equations

$$\begin{cases} J = 0 \\ \frac{\partial J}{\partial z} = 0 \end{cases} \quad (16)$$

We put the origin of the system in the first lens, so we rewrite the lens equation as follows

$$\zeta = z - \frac{1}{(z)^{\frac{n-1}{2}}(\bar{z})^{\frac{n+1}{2}}} - \frac{\gamma^{m+1}}{(z-s)^{\frac{m-1}{2}}(\bar{z}-s)^{\frac{m+1}{2}}} \quad (17)$$

the Jacobian determinant is

$$J = \frac{1}{4}\left[\left(2 + \frac{n-1}{z^{\frac{n+1}{2}}\bar{z}^{\frac{n+1}{2}}} + \frac{(m-1)\gamma^{m+1}}{(z-s)^{\frac{m+1}{2}}(\bar{z}-s)^{\frac{m+1}{2}}}\right)^2 - \left(\frac{n+1}{z^{\frac{n+3}{2}}\bar{z}^{\frac{n-1}{2}}} + \frac{(m+1)\gamma^{m+1}}{(z-s)^{\frac{m+3}{2}}(\bar{z}-s)^{\frac{m-1}{2}}}\right)\left(\frac{n+1}{z^{\frac{n-1}{2}}\bar{z}^{\frac{n+3}{2}}} + \frac{(m+1)\gamma^{m+1}}{(z-s)^{\frac{m+1}{2}}(\bar{z}-s)^{\frac{m+3}{2}}}\right)\right] \quad (18)$$

and

$$\frac{\partial J}{\partial z} = \frac{1}{4} \left\{ \frac{1}{2} \left[\left(\frac{n+1}{z^{\frac{n+3}{2}} \bar{z}^{\frac{n+1}{2}}} + \frac{(m+1)\gamma^{m+1}}{(z-s)^{\frac{m+3}{2}}(\bar{z}-s)^{\frac{m-1}{2}}} \right) \left(\frac{n^2-1}{z^{\frac{n+1}{2}} \bar{z}^{\frac{n+3}{2}}} + \frac{(m^2-1)\gamma^{m+1}}{(z-s)^{\frac{m+1}{2}}(\bar{z}-s)^{\frac{m+3}{2}}} \right) + \right. \right.$$
$$\left. \left(\frac{(n+1)(n+3)}{z^{\frac{n+5}{2}} \bar{z}^{\frac{n-1}{2}}} + \frac{(m+1)(m+3)\gamma^{m+1}}{(z-s)^{\frac{m+5}{2}}(\bar{z}-s)^{\frac{m-1}{2}}} \right) \left(\frac{n+1}{z^{\frac{n-1}{2}} \bar{z}^{\frac{n+3}{2}}} + \frac{(m+1)\gamma^{m+1}}{(z-s)^{\frac{m-1}{2}}(\bar{z}-s)^{\frac{m+3}{2}}} \right) \right] \quad (19)$$
$$\left. - \left(2 + \frac{n-1}{z^{\frac{n+1}{2}} \bar{z}^{\frac{n+1}{2}}} + \frac{(m-1)\gamma^{m+1}}{(z-s)^{\frac{m+1}{2}}(\bar{z}-s)^{\frac{m+1}{2}}} \right) \left(\frac{n^2-1}{z^{\frac{n+3}{2}} \bar{z}^{\frac{n+1}{2}}} + \frac{(m^2-1)\gamma^{m+1}}{(z-s)^{\frac{m+3}{2}}(\bar{z}-s)^{\frac{m+1}{2}}} \right) \right\}$$

Here we show the analytical procedure to find out s_{IW}; the other transition, s_{CI}, is only found numerically.

We require $z = \bar{z}$ because the beak-to-beak singularity for the intermediate-wide transition occurs along the line that joins the two lenses, and we introduce two variables

$$y_1 = \frac{(s-z)^{m+1}}{z^{n+1}}, \quad y_2 = \frac{(s-z)^{m+2}}{z^{n+2}} \quad (20)$$

we replace y_1 in Equation (18), we solve and we get

$$y_1 = \frac{z^{n+1} - 1}{\gamma^{m+1}}, \quad (21)$$

we substitute y_2 in Equation (19), we solve and we find

$$y_2 = \frac{n+1}{(m+1)\gamma^{m+1}}. \quad (22)$$

We use Equations (21) and (22) in Equation (20), we find two new equations and by a combination of them we get a complicated expression for z

$$\frac{\gamma^{\frac{m+1}{n+2}} \left(1 + \gamma^{\frac{n+1}{n+2}}\right)^{m+1} \left[z + \left(\frac{m+1}{n+1}\right)^{\frac{1}{m+2}} \gamma^{\frac{m+1}{m+2}} z^{\frac{n+2}{m+2}}\right]^{n-m}}{\left[z + \left(\frac{m+1}{n+1}\right)^{\frac{1}{m+2}} \gamma^{\frac{m+1}{m+2}} z^{\frac{n+2}{m+2}}\right]^{n+1} - \left(1 + \gamma^{\frac{n+1}{n+2}}\right)^{n+1}} = 1 \quad (23)$$

that we can solve only numerically. We call this numerical solution z_{IW}. Finally we get the value of the intermediate-wide transition for general n and m

$$s_{IW} = z_{IW} + \left(\gamma^{m+1} z_{IW}^{n+2} \frac{m+1}{n+1} \right)^{\frac{1}{m+2}}. \quad (24)$$

In Figure 17, upper panel, we plot the cases with fixed $n = 1$ and variable m (upper curves). The close-intermediate transition s_{CI} is found numerically (lower curves). We can see that the value of s_{IW} increases with γ and that the transition occurs earlier for greater values of m. The value of s_{CI} has a different behavior: first it decreases with increasing γ, with a minimum around $\gamma = 0.5$, and then it starts to grow up again. Also in this case the transition occurs earlier for greater values of m. We remind the reader that we are working in units of the Einstein radius of the first lens.

For the reversed binary case, in Figure 17, lower panel, we plot the cases with fixed $m = 1$ and vary n. We can see that the value of s_{IW} increases with γ similarly to the case with fixed m. For s_{CI} all curves are very closely packed and have a minimum for a value of γ that depends on the specific choice of n. In particular, for $\gamma = \sqrt{0.1}$, corresponding to the situation in Figure 15, the transition occurs for $n = 1$ at smaller separation than for all other curves.

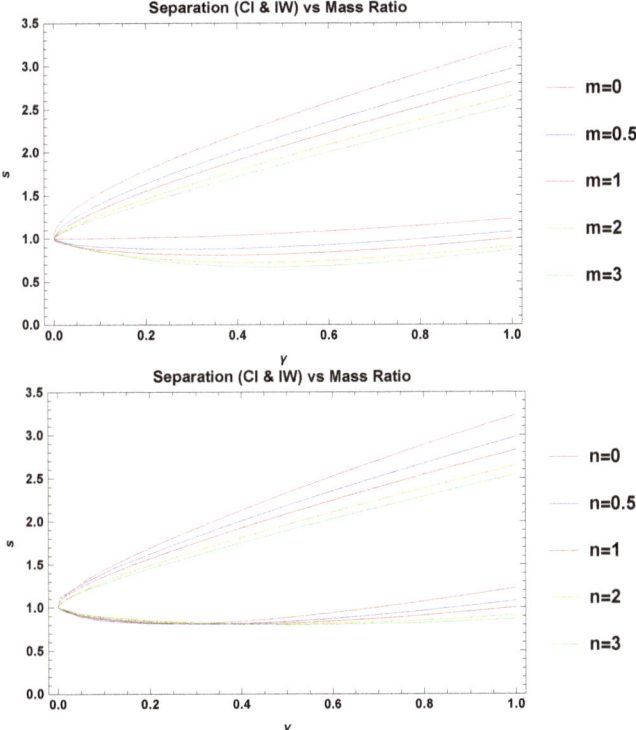

Figure 17. Critical values of the separation for the intermediate-wide transition s_{IW} as a function of γ (upper curves); critical values of the separation for the close-intermediate transition s_{CI} as a function of γ (lower curves). The upper panel is for $n = 1$ and variable m; the lower panel is for $m = 1$ and variable n.

5. Analytical Approximations

In order to remark the differences with the Schwarzschild case $n = m = 1$ and to have a deeper understanding in the caustics evolution, we now want to explore the analytical approximations obtained for the most general case, varying both n and m. In particular, we investigated the wide binary regime and the very small γ regime, but we were not able to get analytical results for the close binary regime. The main difficulty comes from the fact that the starting point of the expansion would have the two lenses coinciding in the origin, but the resulting Einstein radius can only be calculated numerically. Therefore, even the zero order is not analytic.

5.1. Wide Binary

Let us consider the wide binary regime with $s \gg 1$: the case in which an isolated object is perturbed by another one at a distance much greater than the Einstein radius θ_E.

We can set the origin of our system in the first lens ($z_A = 0$ and $z_B = s$), so we can use the same lens equation written in Equation (17) and the Jacobian determinant in Equation (18).

The perturbing object deforms the circular critical curve of the main object with radius ρ and we can find out this deformation δ through a perturbative approach.

We set

$$z = \rho(1+\delta)e^{i\theta} \qquad (25)$$

and we substitute in J. In our case $\rho = 1$ (because for $s \to \infty$ the radius of the critical curves is the Einstein radius, which, in our case, is $\theta_{E,A} = 1$), and take $\delta = O(1/s^{m+1})$. We perform a power series expansion for J about the zero point with respect to $1/s$ at first order. Then, we solve $J = 0$ and find the correction to the critical curve

$$\delta = \frac{\gamma^{m+1}[1 - m + (m+1)\cos(2\theta)]}{2(n+1)s^{m+1}} \tag{26}$$

Now we put Equation (25) in Equation (17) and we expand around zero with respect to $1/s$ at first order, and we take the real and the imaginary parts

$$Re[\zeta(\theta)] = \frac{\gamma^{m+1}}{s^m} + \frac{(m+1)\gamma^{m+1}\cos^3(\theta)}{s^{m+1}} \tag{27}$$

$$Im[\zeta(\theta)] = -\frac{(m+1)\gamma^{m+1}\sin^3(\theta)}{s^{m+1}} \tag{28}$$

The real part contains the shift $\frac{\gamma^{m+1}}{s^m}$ of the caustic toward the direction of perturbing object as we can see in Figures 1, 7, and 12.

The other term, $\cos^3(\theta) + i \sin^3(\theta)$, describes the shape of the caustic (the 4-cusps astroid) that remains unchanged by varying m. The coefficient $\gamma^{m+1}\frac{(m+1)}{s^{m+1}}$ gives the size of the caustic.

In these approximations the next-to-leading order is given by a term $O(1/s^{2(m+1)})$. So, for any values of m the relative importance of these neglected corrections to the analytic caustics of Equations (27) and (28) is $1/s^{m+1}$, which means that for $s = 10$ and $m = 1$ we get a 1% error, while for $m = 0$ we get a 10% error. In fact, caustics with $m = 0$ are more heavily affected by tidal fields.

5.2. Extremely Unequal-Strength Ratio Limit

Now we study the caustic evolution in the extreme limit $\theta_{E,B} \ll \theta_{E,A}$ for the close and wide separations. We remind that, in the case of two Schwarzschild objects ($n = m = 1$) this is the so-called "planetary" limit.

5.2.1. Central Caustic

We put the origin of our system in the first lens ($z_A = 0$), the perturbing object is at $z_B = -s$ and we rewrite the lens equation as follows

$$\zeta = z - \frac{1}{(z)^{\frac{n-1}{2}}(\bar{z})^{\frac{n+1}{2}}} - \frac{\gamma^{m+1}}{(z+s)^{\frac{m-1}{2}}(\bar{z}+s)^{\frac{m+1}{2}}}. \tag{29}$$

The Jacobian determinant is

$$J = \frac{1}{4}\left[\left(2 + \frac{n-1}{z^{\frac{n+1}{2}}\bar{z}^{\frac{n+1}{2}}} + \frac{(m-1)\gamma^{m+1}}{(z+s)^{\frac{m+1}{2}}(\bar{z}+s)^{\frac{m+1}{2}}}\right)^2 - \left(\frac{n+1}{z^{\frac{n+3}{2}}\bar{z}^{\frac{n-1}{2}}} + \frac{(m+1)\gamma^{m+1}}{(z+s)^{\frac{m+3}{2}}(\bar{z}+s)^{\frac{m-1}{2}}}\right)\left(\frac{n+1}{z^{\frac{n-1}{2}}\bar{z}^{\frac{n+3}{2}}} + \frac{(m+1)\gamma^{m+1}}{(z+s)^{\frac{m-1}{2}}(\bar{z}+s)^{\frac{m+3}{2}}}\right)\right] \tag{30}$$

For the critical curve of the main lens, we use the parameterization in J

$$z = (1+\delta)e^{i\theta} \tag{31}$$

where $\delta = O(\gamma^{m+1})$.

We expand around zero with respect to γ^{m+1} to first order, solve $J = 0$ and we find the correction of the circular critical curve

$$\delta = \frac{2 + 4s\cos\theta + s^2[(m+1)\cos(2\theta) - m + 1]}{2(n+1)\left(1 + 2s\cos\theta + s^2\right)^{\frac{m+3}{2}}}\gamma^{m+1}. \tag{32}$$

We can substitute this δ in Equation (29) and obtain the caustic. Since it is not a simple expression, we omit it here.

In order to find the size of the caustic, we evaluate it for $\theta = 0$ and $\theta = \pi$. We have

$$\Delta\zeta = \zeta(0) - \zeta(\pi) = s\gamma^{m+1}\left[\frac{1}{(s-1)^{m+1}} - \frac{1}{(s+1)^{m+1}}\right] \tag{33}$$

and this is the distance between the left and the right cusp.

We also find that the caustic is invariant under the transformation

$$s \to \frac{1}{s}, \gamma^{m+1} \to \frac{\gamma^{m+1}}{s^{m-1}}. \tag{34}$$

which expresses the duality of the close-wide regimes in our mixed binary framework.

5.2.2. Caustics of the Perturbing Object

We put the origin of our system in the second lens so that $z_A = -s$ and $z_B = 0$ and the lens equation becomes

$$\zeta = z - \frac{1}{(z+s)^{\frac{n-1}{2}}(\bar{z}+s)^{\frac{n+1}{2}}} - \frac{\gamma^{m+1}}{z^{\frac{m-1}{2}}\bar{z}^{\frac{m+1}{2}}} \tag{35}$$

we rewrite the Jacobian determinant as follows

$$J = \frac{1}{4}\left[\left(2 + \frac{n-1}{(z+s)^{\frac{n+1}{2}}(\bar{z}+s)^{\frac{n+1}{2}}} + \frac{(m-1)\gamma^{m+1}}{(z)^{\frac{m+1}{2}}(\bar{z})^{\frac{m+1}{2}}}\right)^2 - \left(\frac{n+1}{(z+s)^{\frac{n+3}{2}}(\bar{z}+s)^{\frac{n-1}{2}}} + \frac{(m+1)\gamma^{m+1}}{z^{\frac{m+3}{2}}\bar{z}^{\frac{m-1}{2}}}\right)\left(\frac{n+1}{(z+s)^{\frac{n-1}{2}}(\bar{z}+s)^{\frac{n+3}{2}}} + \frac{(m+1)\gamma^{m+1}}{z^{\frac{m-1}{2}}\bar{z}^{\frac{m+3}{2}}}\right)\right] \tag{36}$$

and we introduce a new expression for z

$$z = \rho^{\frac{1}{m+1}}\gamma e^{i\theta}. \tag{37}$$

We substitute in Equation (36) and we expand with respect to γ^{m+1}, around zero at zero order and the Jacobian determinant becomes

$$\frac{(\rho-1)(\rho+m)}{\rho^2} + \frac{(n-1)(m+2\rho-1) - (m+1)(n+1)\cos(2\theta)}{2\rho s^2} - \frac{n}{s^{2n+2}} = 0 \tag{38}$$

Then we solve Equation (38), $J = 0$, with respect to ρ and we find two solutions

$$\rho_\pm = \frac{(m-1)\{s^{n+1}[(1-n) + (m+1)(n+1)\cos(2\theta) - 2s^{n+1}] \pm \sqrt{\Delta}\}}{4\left(s^{n+1} - 1\right)\left(s^{n+1} + n\right)} \tag{39}$$

$$\Delta = s^{2n+2}\{[(m-1)(2s^{n+1} + n - 1) - (m+1)(n+1)\cos(2\theta)]^2 + 16m(s^{n+1} - 1)(s^{n+1} + n)\} \tag{40}$$

We have two scenarios: for external objects (when the secondary lens is outside the Einstein ring of the main lens, $s > 1$) the critical curves are elongated rings, see Figure 12; for internal objects (when the secondary lens is inside the critical curve of the main lens, $s < 1$) it generates two specular ovals, see Figure 16.

In order to get the caustics we put our solutions in the lens equation and we get, at first order:

$$\zeta = \gamma \rho^{\frac{1}{m+1}} \left[e^{i\theta} \left(\frac{n-1}{2s^{n+1}} - \frac{1}{\rho} + 1 \right) + \frac{e^{-i\theta}(n+1)}{2s^{n+1}} \right] - \frac{1}{s^n} \quad (41)$$

From Equation (41) we can get all the information for the size and for the displacement of the secondary caustic from the central one.

The displacement along the axis that joins the two lenses is the middle point $[\zeta(0) + \zeta(\pi)]/2$ and in our case is

$$\zeta_{center} = s - \frac{1}{s^n}. \quad (42)$$

because the origin of our system is in the second lens.

Now we want to find out the size of the caustics in the close and wide separation.

For the wide case we have an extension of the caustics in the parallel direction (with respect to the lens axis), given by $[\zeta(0) - \zeta(\pi)]$

$$\Delta \zeta_{\parallel,wide} = 2(n+1) \frac{\gamma}{s^{\frac{m(n+1)}{m+1}} (s^{n+1} - 1)^{\frac{1}{m+1}}} \quad (43)$$

and in the vertical direction, orthogonal to lens axis, $[\zeta(-\pi/2) - \zeta(\pi/2)]$:

$$\Delta \zeta_{\perp,wide} = 2(n+1) \frac{\gamma}{s^{\frac{m(n+1)}{m+1}} (s^{n+1} + n)^{\frac{1}{m+1}}} \quad (44)$$

In Figure 18, upper panel, we show the size of the caustic for three different fixed $n = 0.5, 1, 2$ with variable m. Keeping γ and s fixed, the size is almost independent of m, as can be seen by neglecting n in the sum in the denominator of Equation (43).

In the close regime, in order to find the position of the central caustic, we evaluate the lens equation in $\rho_\pm$ for $\theta = \pm \pi/2$ and we need to distinguish the case $mn > 1$ from the case $mn < 1$ changing the sign after taking the square root.

We put $\theta = \pi/2$ and ρ_+ in Equation (41), we take the imaginary part and we find

$$Im[\zeta_+(\pi/2)] = \begin{cases} (m+1)\gamma \left[\frac{(1-s^{n+1})}{ms^{n+1}} \right]^{\frac{m}{m+1}}, & \text{if } mn < 1 \\ (n+1)\gamma \left[\frac{1}{s^{m(n+1)}(n+s^{n+1})} \right]^{\frac{1}{m+1}}, & \text{if } mn > 1 \end{cases} \quad (45)$$

In order to find the position of the two secondary caustics we need to calculate $\zeta(\pi/2)$ with ρ_-, then we take the imaginary part and we find

$$Im[\zeta_-(\pi/2)] = \begin{cases} (n+1)\gamma \left[\frac{1}{s^{m(n+1)}(n+s^{n+1})} \right]^{\frac{1}{m+1}}, & \text{if } mn < 1 \\ (m+1)\gamma \left[\frac{(1-s^{n+1})}{ms^{n+1}} \right]^{\frac{m}{m+1}}, & \text{if } mn > 1 \end{cases} \quad (46)$$

The measure of the transverse size of the secondary caustics is the difference between the last two formulas, $Im[\zeta_+(\pi/2)] - Im[\zeta_-(\pi/2)]$ and we plot the result in Figure 18, lower panel, for $s = 0.6$, $\gamma = 0.01$, for three different fixed $n = 0.5, 1, 2$, with variable m.

We can see that the size increases with m and n, coherently with what is found in [58], where large values of m and n produce giant triangular caustics in the close regime. We finally stress that, for $mn < 1$, and so for $n = 0.5$ especially (green line), the two branches exchange role because of the elliptic umbilic catastrophe. Then, we must change the sign in our formula for the size.

In these approximations the error is given by a term $O(\gamma^2)$, which means a 1% error for the case $\gamma = 0.01$ examined in our plots.

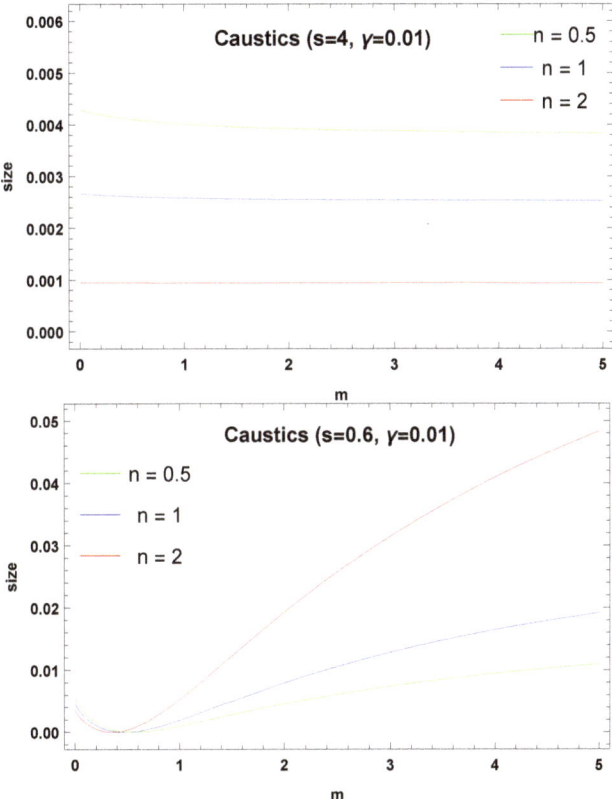

Figure 18. Upper panel: size of the caustic in the wide case for $s = 4$, $\gamma = 0.01$, for three different fixed $n = 0.5, 1, 2$, with variable m. Lower panel: size of the caustic in the close case for $s = 0.6$, $\gamma = 0.01$, for three different fixed $n = 0.5, 1, 2$, with variable m.

6. Conclusions

In this paper we have generalized our previous study of binary lenses with $1/r^n$ potential [58], by extending it to the case of mixed binaries. Of course the mathematics of this general case is interesting from several points of view, since many earlier results are put in a more general context. However, this case is also important from the astrophysical point of view. In fact, we now have the critical curves and caustics of pairs of galaxies with different halos, or we may apply our results to cases in which one object is made up of exotic matter and the other one is a normal star. For direct applications of our results to astrophysical objects, we remind that all plots are in units of the Einstein radius of the first lens. This can be calculated by standard formulae for any specific lens models.

Our figures, together with those of [58] may be considered as a complete atlas of critical curves and caustics in binary lensing by $1/r^n$ potentials. We studied different limits in which the stronger (weaker) lens has a steeper (gentler) potential in all three topology regimes.

We have shown that an elliptic umbilic catastrophe exists for $mn < 1$ and calculated its position. We have also described the pseudocaustic in the $m = 0$ limit. We have calculated the boundaries of the three topology regimes and provided analytic approximations for the wide binary and the extremely small-strength secondary lens.

With respect to the $m=n$ binary lens case, we note that for large m we still have large secondary triangular caustics, but they are not as giant as those in [58]. In fact, the presence of a more standard lens in the system mitigates the behavior at large distances and pushes back these caustics to more normal sizes. Indeed, these structures are quite sensitive to the parameters of the lens.

This fact helps us recall that the mixed binary lens described here is still obtained by the linear superposition of the potentials of two isolated objects. This is physically relevant whenever we can neglect the non-linear terms in Einstein equations. Even when this is not possible, our results may serve as a basis for more accurate calculations.

Author Contributions: Supervision V.B., investigation S.P., methodology C.M. All authors have read and agreed to the published version of the manuscript.

Funding: This research received no external funding.

Conflicts of Interest: The authors declare no conflict of interest.

References

1. Einstein, A.; Grossmann, M. *Entwurf einer Verallgemeinerten Relativitätstheorie und eine Theorie der Gravitation*; Teubner: Leipzig/Berlin, Germany, 1913; Volume 6, pp. 225–261.
2. Dyson, F.W.; Eddington, A.S.; Davidson, C. A Determination of the deflection of light by the sun's gravitational field, from observations made at the total eclipse of 29 May 1919. *Philos. Trans. R. Soc. Lond. Ser. A* **1920**, *220*, 291–333.
3. Gaudi, B.S. Microlensing Surveys for Exoplanets. *Ann. Rev. Astr. Astroph.* **2012**, *50*, 411. [CrossRef]
4. Niikura, H.; Takada, M.; Yasuda, N.; Lupton, R.H.; Sumi, T.; More, S.; Kurita, T.; Sugiyama, S.; More, A.; Oguri, M.; et al. Microlensing constraints on primordial black holes with Subaru/HSC Andromeda observations. *Nat. Astr.* **2019**, *3*, 524. [CrossRef]
5. Gavazzi, R.; Treu, T.; Rhodes, J.D.; Koopmans, L.V.; Bolton, A.S.; Burles, S.; Massey, R.J.; Moustakas, L.A. The Sloan Lens ACS Survey. IV: The mass density profile of early-type galaxies out to 100 effective radii. *Astrophys. J.* **2007**, *667*, 176–190. [CrossRef]
6. Hoekstra, H.; Jain, B. Weak Gravitational Lensing and Its Cosmological Applications. *Ann. Rev. Nucl. Part. Sci.* **2008**, *58*, 99–123. [CrossRef]
7. Hoekstra, H.; Bartelmann, M.; Dahle, H.; Israel, H.; Limousin, M.; Meneghetti, M. Masses of galaxy clusters from gravitational lensing. *Space Sci. Rev.* **2013**, *177*, 75–118. [CrossRef]
8. Kiblinger, M. Cosmology with cosmic shear observations: A review. *Rep. Prog. Phys.* **2015**, *78*, 086901.
9. Jeong, D.; Schmidt, F. Large-scale structure observables in general relativity. *Class. Quant. Gravity* **2015**, *32*, 044001. [CrossRef]
10. Schneider, P.; Ehlers, J.; Falco, E.E. *Gravitational Lenses*; Springer: Berlin/Heidelberg, Germany; New York, NY, USA, 1992.
11. Petters, A.O.; Levine, H.; Wambsganss, J. *Singularity Theory and Gravitational Lensing*; Progress in Mathematical Physics; Springer Science+, Birkhauser: Basel, Switzerland, 2001; Volume XXV, p. 603.
12. Perlick, V. Gravitational Lensing from a Spacetime Perspective. *Living Rev. Relativ.* **2004**, *7*, 9. [CrossRef]
13. Schneider, P.; Kochanek, C.S.; Wambsganss, J. *Gravitational Lensing: Strong, Weak and Micro*; Lecture Notes of the 33rd Saas-Fee Advanced Course; Springer: Berlin, Germany, 2006; Volume XVI, p. 552.
14. Zakharov, A.F. Lensing by exotic objects. *Gen. Relativ. Gravit.* **2010**, *42*, 2301. [CrossRef]
15. Bozza, V. Gravitational Lensing by Black Holes. *Gen. Relativ. Gravit.* **2010**, *42*, 2269–2300. [CrossRef]
16. Kim, S.W.; Cho, Y.M. Wormhole gravitational lens. In *Evolution of the Universe and Its Observational Quest*; Universal Academy Press: Tokyo, Japan, 1994; pp. 353–354.
17. Cramer, J.G.; Forward, R.L.; Morris, M.S.; Visser, M.; Benford, G.; Landis, G.A. Natural Wormholes as Gravitational Lenses. *Phys. Rev. D* **1995**, *51*, 3117. [CrossRef] [PubMed]
18. Safonova, M.; Torres, D.F.; Romero, G.E. Macrolensing signatures of large-scale violations of the weak energy condition. *Mod. Phys. Lett. A* **2001**, *16*, 153. [CrossRef]

19. Eiroa, E.; Romero, G.E.; Torres, D.F. Chromaticity effects in microlensing by wormholes. *Mod. Phys. Lett. A* **2001**, *16*, 973. [CrossRef]
20. Safonova, M.; Torres, D.F.; Romero, G.E. Microlensing by natural wormholes: Theory and simulations. *Phys. Rev. D* **2001**, *65*, 023001. [CrossRef]
21. Safonova, M.; Torres, D.F. Degeneracy in exotic gravitational lensing. *Mod. Phys. Lett. A* **2002**, *17*, 1685. [CrossRef]
22. Takahashi, R.; Asada, H. Observational Upper Bound on the Cosmic Abundances of Negative-mass Compact Objects and Ellis Wormholes from the Sloan Digital Sky Survey Quasar Lens Search. *Astrophys. J.* **2013**, *768*, L16. [CrossRef]
23. Rahaman, F.; Kalam, M.; Chakraborty, S. Gravitational Lensing by a Stable C-Field Wormhole. *Chin. J. Phys.* **2007**, *45*, 518.
24. Kuhfittig, P.K.F. Gravitational lensing of wormholes in noncommutative geometry. *arXiv* **2015**, arXiv:1501.06085.
25. Tejeiro, S.; Larranaga, R. Gravitational lensing by wormholes. *Rom. J. Phys.* **2012**, *57*, 736.
26. Nandi, K.K.; Zhang, Y.-Z.; Zakharov, A.V. Gravitational lensing by wormholes. *Phys. Rev. D* **2006**, *74*, 024020. [CrossRef]
27. Dey, T.K.; Sen, S. Gravitational lensing by wormholes. *Mod. Phys. Lett. A* **2008**, *23*, 953. [CrossRef]
28. Ellis, H.G. Ether flow through a drainhole: A particle model in general relativity. *J. Math. Phys.* **1973**, *14*, 104. [CrossRef]
29. Bronnikov, K.A. Scalar-tensor theory and scalar charge. *Acta Phys. Pol. B* **1973**, *4*, 251–266.
30. Morris, M.S.; Thorne, K.S. Wormholes in spacetime and their use for interstellar travel: A tool for teaching general relativity. *Am. J. Phys.* **1988**, *56*, 395. [CrossRef]
31. Morris, M.S.; Thorne, K.S.; Yurtsever, U. Wormholes, Time Machines, and the Weak Energy Condition. *Phys. Rev. Lett.* **1988**, *61*, 1446. [CrossRef]
32. Teo, E. Rotating traversable wormholes. *Phys. Rev. D* **1998**, *58*, 024014. [CrossRef]
33. Jusufi, K.; Ovgun, A. Gravitational Lensing by Rotating Wormholes. *Phys. Rev. D* **2018**, *97*, 024042. [CrossRef]
34. Tsukamoto, N.; Harada, T. Light curves of light rays passing through a wormhole. *Phys. Rev. D* **2017**, *95*, 024030. [CrossRef]
35. Ohgami, T.; Sakai, N. Wormhole Shadows. *Phys. Rev. D* **2015**, *91*, 124020. [CrossRef]
36. Ohgami, T.; Sakai, N. Wormhole Shadows in Rotating Dust. *Phys. Rev. D* **2016**, *94*, 064071. [CrossRef]
37. Chetouani, L.; Clément, G. Geometrical optics in the Ellis geometry. *Gen. Relat. Gravit.* **1982**, *16*, 111–119. [CrossRef]
38. Perlick, V. Exact gravitational lens equation in spherically symmetric and static spacetimes. *Phys. Rev. D* **2004**, *69*, 064017. [CrossRef]
39. Abe, F. Gravitational Microlensing by the Ellis Wormhole. *Astrophys. J.* **2010**, *725*, 787. [CrossRef]
40. Bhattacharya, A.; Potapov, A.A. Bending of light in Ellis wormhole geometry. *Mod. Phys. Lett. A* **2010**, *25*, 2399. [CrossRef]
41. Toki, Y.; Kitamura, T.; Asada, H.; Abe, F. Astrometric Image Centroid Displacements due to Gravitational Microlensing by the Ellis Wormhole. *Astrophys. J.* **2011**, *740*, 121. [CrossRef]
42. Tsukamoto, N.; Harada, T.; Yajima, K. Can we distinguish between black holes and wormholes by their Einstein-ring systems? *Phys. Rev. D* **2012**, *86*, 104062. [CrossRef]
43. Nakajima, K.; Asada, H. Deflection angle of light in an Ellis wormhole geometry. *Phys. Rev. D* **2012**, *85*, 107501. [CrossRef]
44. Gibbons, G.W.; Vyska, M. The Application of Weierstrass elliptic functions to Schwarzschild Null Geodesics. *Class. Quant. Grav.* **2012**, *29*, 065016. [CrossRef]
45. Yoo, C.M.; Harada, T.; Tsukamoto, N. Wave Effect in Gravitational Lensing by the Ellis Wormhole. *Phys. Rev. D* **2013**, *87*, 084045. [CrossRef]
46. Kitamura, T.; Nakajima, K.; Asada, H. Demagnifying gravitational lenses toward hunting a clue of exotic matter and energy. *Phys. Rev. D* **2013**, *87*, 027501. [CrossRef]
47. Izumi, K.; Hagiwara, C.; Nakajima, K.; Kitamura, T.; Asada, H. Gravitational lensing shear by an exotic lens object with negative convergence or negative mass. *Phys. Rev. D* **2013**, *88*, 024049. [CrossRef]
48. Tsukamoto, N.; Kitamura, T.; Nakajima, K.; Asada, H. Gravitational lensing in Tangherlini spacetime in the weak gravitational field and the strong gravitational field. *Phys. Rev. D* **2014**, *90*, 064043. [CrossRef]

49. Nakajima, K.; Izumi, K.; Asada, H. Negative time delay of light by a gravitational concave lens. *Phys. Rev. D* **2014**, *90*, 084026. [CrossRef]
50. Tsukamoto, N.; Harada, T. Signed magnification sums for general spherical lenses. *Phys. Rev. D* **2013**, *87*, 024024. [CrossRef]
51. Kitamura, T. Microlensed image centroid motions by an exotic lens object with negative convergence or negative mass. *Phys. Rev. D* **2014**, *89*, 084020. [CrossRef]
52. Bozza, V.; Postiglione, A. Alternatives to Schwarzschild in the weak field limit of General Relativity. *J. Cosmol. Astropart. Phys.* **2015**, *06*, 036. [CrossRef]
53. Bisnovatyi-Kogan, G.S.; Tsupko, O.Y. Gravitational Lensing in a Non-Uniform Plasma. *Mon. Not. R. Astron. Soc.* **2010**, *404*, 1790–1800. [CrossRef]
54. Xinzhong, E.; Rogers, A. Two Families of Astrophysical Diverging Lens Models. *Mon. Not. R. Astron. Soc.* **2017**, *475*, 867–878.
55. Xinzhong, E.; Rogers, A. Dual-Component Plasma Lens Models. *Mon. Not. R. Astron. Soc.* **2019**, *485*, 5800–5816.
56. Xinzhong, E.; Rogers, A. Two Families of Elliptical Plasma Lenses. *Mon. Not. R. Astron. Soc.* **2019**, *488*, 5651–5664.
57. Bisnovatyi-Kogan, G.S.; Tsupko, O.Y. Hills and Holes in the Microlensing Light Curve Due to Plasma Environment Around Gravitational Lens. *Mon. Not. R. Astron. Soc.* **2019**, *491*, 5636–5649.
58. Bozza, V.; Melchiorre, C. Caustics of $1/r^n$ binary gravitational lenses: From galactic haloes to exotic matter. *J. Cosmol. Astropart. Phys.* **2016**, *03*, 040. [CrossRef]
59. Asada, H. Gravitational lensing by exotic objects. *Mod. Phys. Lett. A* **2017**, *32*, 1730031. [CrossRef]
60. Akiyama, K.; Alberdi, A.; Alef, W.; Asada, K.; Azulay, R.; Baczko, A.K.; Ball, D.; Baloković, M.; Barrett, J.; Bintley, D.; et al. [Event Horizon Telescope Collaboration]. First M87 Event Horizon Telescope Results. IV. Imaging the Central Supermassive Black Hole. *Astrophys. J.* **2019**, *875*, L1.
61. Tsukamoto N.; Kokubu, T. High energy particle collisions in static, spherically symmetric black-hole-like wormholes. *Phys. Rev. D* **2020**, *101*, 044030. [CrossRef]
62. Damour, T.; Solodukhin, S.N. Wormholes as black hole foils. *Phys. Rev. D* **2007**, *76*, 024016. [CrossRef]
63. Shin, E.M.; Evans, N.W. Lensing by binary galaxies modelled as isothermal spheres. *Mon. Not. R. Astron. Soc.* **2008**, *390*, 505. [CrossRef]
64. Kovner, I. The Quadrupole Gravitational Lens. *Astrophys. J.* **1987**, *312*, 22. [CrossRef]
65. Evans, N.W.; Wilkinson, M. Lens Models with Density Cusps. *Mon. Not. R. Astron. Soc.* **1998**, *296*, 800. [CrossRef]
66. Rhie, S.H. Elliptically Symmetric Lenses and Violation of Burke's Theorem. *arXiv* **2010**, arXiv:1006.0782.
67. Wang, Y.; Turner, E.L. Caustics, critical curves and cross-sections for gravitational lensing by disc galaxies. *Mon. Not. R. Astron. Soc.* **1997**, *292*, 863. [CrossRef]
68. Tessore, N.; Metcalf, R.B. The elliptical power law profile lens. *Astron. Astrophys.* **2015**, *580*, A79. [CrossRef]
69. Lake, E.; Zheng, Z. Gravitational lensing by ring-like structures. *Mon. Not. R. Astron. Soc.* **2016**, *465*, 2018–2032. [CrossRef]
70. Aazami, A.; Keeton, C.; Petters, A. Magnification Cross Sections for the Elliptic Umbilic Caustic Surface. *Universe* **2019**, *5*, 161. [CrossRef]
71. Witt, H.J. Investigation of high amplification events in light curves of gravitationally lensed quasars. *Astron. Astrophys.* **1990**, *236*, 311.
72. Schneider, P.; Weiss, A. The two-point-mass lens-Detailed investigation of a special asymmetric gravitational lens. *Astron. Astrophys.* **1986**, *164*, 237.
73. Erdl, H.; Schneider, P. Classification of the multiple deflection two point-mass gravitational lens models and application of catastrophe theory in lensing. *Astron. Astrophys.* **1993**, *268*, 453.

© 2020 by the authors. Licensee MDPI, Basel, Switzerland. This article is an open access article distributed under the terms and conditions of the Creative Commons Attribution (CC BY) license (http://creativecommons.org/licenses/by/4.0/).

Review

The Effects of Finite Distance on the Gravitational Deflection Angle of Light

Toshiaki Ono and Hideki Asada *

Graduate School of Science and Technology, Hirosaki University, Aomori 036-8561, Japan; ono@tap.st.hirosaki-u.ac.jp
* Correspondence: asada@hirosaki-u.ac.jp

Received: 19 July 2019; Accepted: 21 October 2019; Published: 1 November 2019

Abstract: In order to clarify the effects of the finite distance from a lens object to a light source and a receiver, the gravitational deflection of light has been recently reexamined by using the Gauss–Bonnet (GB) theorem in differential geometry (Ishihara et al. 2016). The purpose of the present paper is to give a short review of a series of works initiated by the above paper. First, we provide the definition of the gravitational deflection angle of light for the finite-distance source and receiver in a static, spherically symmetric and asymptotically flat spacetime. We discuss the geometrical invariance of the definition by using the GB theorem. The present definition is used to discuss finite-distance effects on the light deflection in Schwarzschild spacetime for both the cases of weak deflection and strong deflection. Next, we extend the definition to stationary and axisymmetric spacetimes. We compute finite-distance effects on the deflection angle of light for Kerr black holes and rotating Teo wormholes. Our results are consistent with the previous works if we take the infinite-distance limit. We briefly mention also the finite-distance effects on the light deflection by Sagittarius A*.

Keywords: gravitational lens; general relativity; black hole; wormhole

1. Introduction

In 1919, the experimental confirmation of the theory of general relativity [1] succeeded [2]. It is the measurement of the gravitational deflection angle of light. Since then, the gravitational deflection angle of light has attracted a lot of attention. Many authors have studied the gravitational deflection of light by black holes [3–16]. The gravitational lens by other objects such as wormholes and gravitational monopoles also has attracted a lot of interest [17–30]. Very recently, the Event Horizon Telescope (EHT) team has reported a direct image of the inner edge of the hot matter around the black hole candidate at the center of M87 galaxy [31–36]. The direct imaging of black hole shadows must again and steeply raise the importance of the gravitational deflection angle of light.

Most of those calculations are based on the coordinate angle. The angle respects the rotational symmetry of the spacetime. Gibbons and Werner (2008) made an attempt at defining, in a more geometrical manner, the deflection angle of light [37]. In their paper, the source and receiver are needed to be located at an asymptotic Minkowskian region. The Gauss–Bonnet theorem was applied to a spatial domain by introducing the optical metric, for which a light ray is expressed as a spatial geodesic curve. Ishihara et al. have successfully extended Gibbons and Werner's idea such that the source and receiver can be at a finite distance from the lens object [38]. They extend the earlier work to the case of the strong deflection limit, in which the winding number of the photon orbits may be larger than unity [39]. In particular, the asymptotic receiver and source are not needed. Arakida [40] made an attempt to apply the Gauss–Bonnet theorem to quadrilaterals that are not extending to infinity and proposed a new definition of the deflection angle of light, though a comparison between two different manifolds that he proposed is an open issue. Proposing an alternative definition of the deflection angle of light, Crisnejo et al. [41] has recently made a comparison between the alternative definitions in References [38–40] and showed by explicit calculations that the definition by Arakida in Reference [40] is different from that by Ishihara et al. [38,39]. Their definition has been applied to study gravitational lensing with a plasma medium [41].

The earlier works [38,39] are restricted within the spherical symmetry. Ono et al. have extended the Gauss–Bonnet method with the optical metric to axisymmetric spacetimes [42]. This extension includes mathematical quantities and calculations, with which most physicists are not very familiar. Therefore, the purpose of this paper provides a review of the series of papers on the gravitational deflection of light for finite-distance sources and receivers. In particular, we hope that the detailed calculations in this paper will be helpful for readers to compute the gravitational deflection of light by the new powerful method. For instance, this new technique has been used to study the gravitational lensing in rotating Teo wormholes [43] and also in Damour–Solodukhin wormholes [44]. This formulation has been successfully used to clarify the deflection of light in a rotating global monopole spacetime with a deficit angle [45].

This paper is organized as follows. Section 2 discusses the definition of the gravitational deflection angle of light in static and spherically symmetric spacetimes. Section 3 considers the weak deflection of light in Schwarzschild spacetime. Section 4 discusses the weak deflection of light in the Kottler spacetime and the Weyl conformal gravity model. The strong deflection of light is examined in Section 5. Sagittarius A* (Sgr A*) is also discussed as an example for possible candidates. In Section 6, we discuss the strong deflection of light with finite-distance corrections in Schwarzschild spacetime. Section 7 proposes the definition of the gravitational deflection angle of light in stationary and axisymmetric spacetimes. Sgr A* is also discussed. The weak deflection of light is discussed for Kerr spacetime in Section 8 and for rotating Teo wormholes in Section 9. Section 10 is a summary of this paper. Appendix A provides the detailed calculations for the Kerr spacetime. Throughout this paper, we use the unit of $G = c = 1$ and the observer may be called the receiver in order to avoid confusion between r_O and r_0 by using r_R.

2. Definition of the Gravitational Deflection Angle of Light: Static and Spherically Symmetric Spacetimes

Notation

Following Ishihara et al. [38], this section begins by considering a static and spherically symmetric (SSS) spacetime. The metric of this spacetime can be written as:

$$\begin{aligned} ds^2 &= g_{\mu\nu}dx^\mu dx^\nu \\ &= -A(r)dt^2 + B(r)dr^2 + r^2 d\Omega^2, \end{aligned} \quad (1)$$

where $d\Omega^2 \equiv d\theta^2 + \sin^2\theta d\phi^2$ and t, θ, and ϕ are associated with the symmetries of the SSS spacetime. For a metric of the form in Equation (1), we always have to restrict to the domain where $A(r)$ and $B(r)$ are positive such that a static emitter and a static receiver can exist. The spacetime has a spherical symmetry. Therefore, the photon orbital plane is chosen without loss of generality as the equatorial plane ($\theta = \pi/2$). We follow the usual definition of the impact parameter of the light ray as:

$$\begin{aligned} b &\equiv \frac{L}{E} \\ &= \frac{r^2}{A(r)} \frac{d\phi}{dt}. \end{aligned} \quad (2)$$

From $ds^2 = 0$ for the light ray, the orbit Equation is derived as:

$$\left(\frac{dr}{d\phi}\right)^2 + \frac{r^2}{B(r)} = \frac{r^4}{b^2 A(r) B(r)}. \quad (3)$$

Light rays are described by the null condition $ds^2 = 0$, which is solved for dt^2 as:

$$dt^2 = \gamma_{IJ} dx^I dx^J$$
$$= \frac{B(r)}{A(r)} dr^2 + \frac{r^2}{A(r)} d\phi^2, \qquad (4)$$

where I and J denote 1 and 2 and we used Equation (1). We refer to γ_{IJ} as the optical metric. The optical metric can be used to describe a two-dimensional Riemannian space. This Riemannian space is denoted as M^{opt}. The light ray is a spatial geodetic curve in M^{opt}.

In the optical metric space M^{opt}, let Ψ denote the angle between the light propagation direction and the radial direction. A straightforward calculation gives:

$$\cos \Psi = \frac{b\sqrt{A(r)B(r)}}{r^2} \frac{dr}{d\phi}. \qquad (5)$$

This is rewritten as:

$$\sin \Psi = \frac{b\sqrt{A(r)}}{r}, \qquad (6)$$

where we used Equation (3).

We denote Ψ_R and Ψ_S as the directional angles of light propagation. Ψ_R and Ψ_S are measured at the receiver position (R) and the source position (S), respectively. We denote $\phi_{RS} \equiv \phi_R - \phi_S$ as the coordinate separation angle between the receiver and source. By using angles Ψ_R, Ψ_S, and ϕ_{RS}, we define the following:

$$\alpha \equiv \Psi_R - \Psi_S + \phi_{RS}. \qquad (7)$$

This is a basic tool that was invented in Reference [38]. In the following, we shall prove that the definition by Equation (7) is geometrically invariant [38,39].

Here, we briefly mention the Gauss–Bonnet theorem. T is a two-dimensional orientable surface. Differentiable curves ∂T_a ($a = 1, 2, \cdots, N$) are its boundaries. Please see Figure 1 for the orientable surface. We denote the jump angles between the curves as θ_a ($a = 1, 2, \cdots, N$). The Gauss–Bonnet theorem is as follows [46]:

$$\iint_T K dS + \sum_{a=1}^N \int_{\partial T_a} \kappa_g d\ell + \sum_{a=1}^N \theta_a = 2\pi, \qquad (8)$$

where ℓ means the line element of the boundary curve, dS denotes the area element of the surface, K means the Gaussian curvature of the surface T, and κ_g is the geodesic curvature of ∂T_a. The sign of ℓ is chosen to be consistent with the surface orientation.

Suppose a quadrilateral ${}_R^\infty \square_S^\infty$. Please see Figure 2 for this. This is made of four lines: (1) the spatial curve for the light ray, (2, 3) two outgoing radial lines from R and from S, and (4) a circular arc segment C_r that is centered at the lens with the coordinate radius r_C ($r_C \to \infty$) and intersects the radial lines at the receiver or the source. We restrict ourselves within the asymptotically flat spacetime. Then, $\kappa_g \to 1/r_C$ and $d\ell \to r_C d\phi$ as $r_C \to \infty$ (See, e.g., Reference [37]). By using them, we find $\int_{C_r} \kappa_g d\ell \to \phi_{RS}$. Applying this result to the Gauss–Bonnet theorem for ${}_R^\infty \square_S^\infty$, we obtain:

$$\alpha = \Psi_R - \Psi_S + \phi_{RS}$$
$$= -\iint_{{}_R^\infty \square_S^\infty} K dS. \qquad (9)$$

Therefore, α is shown to be invariant for transformations of the spatial coordinates. In addition, α is well defined even when L is a singular point. This is because the point L does not appear in the surface integral nor in the line integral. Furthermore, α vanishes in Euclidean space. This means α is a measure of the deviation from the flat space.

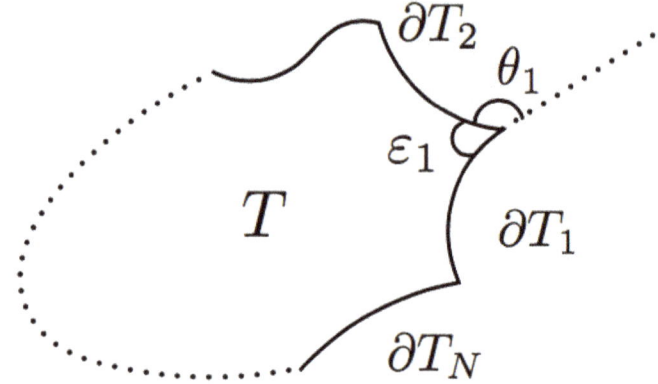

Figure 1. Gauss–Bonnet theorem: We consider a closed curve in a surface.

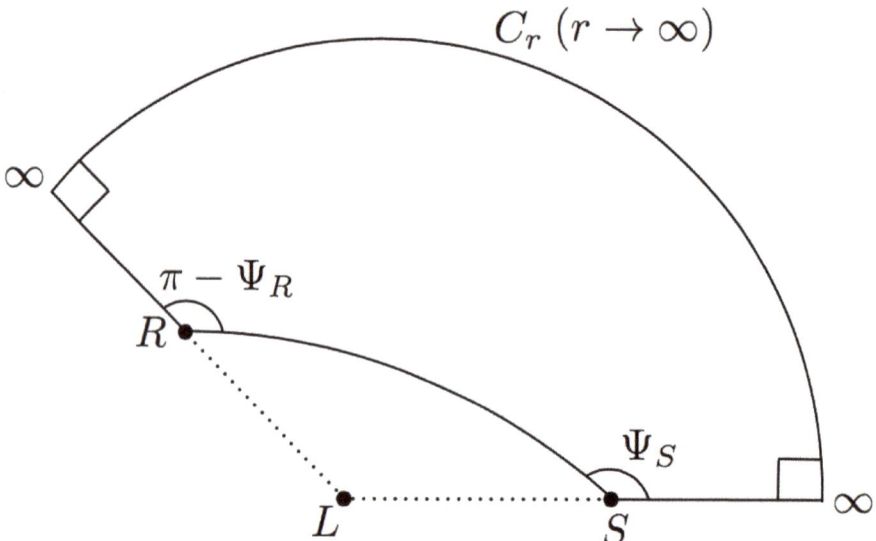

Figure 2. $_R^\infty\square_S^\infty$ is a quadrilateral embedded in a curved space.

Here, we explain that α defined in Equation (7) is observable in principle. For simplicity, let us imagine the following ideal situation. The positions of a source and a receiver are known. For instance, we assume that the lens object is the Sun, the receiver is located at the Earth, and the source is a pulsar which radiates radio signals with a constant period in an anisotropic manner. In particular, we assume that the source is one of the known pulsars of which the spin period and pulse signal behaviors such as pulse profiles are well understood. By very accurate radio observations such as Very Long Baseline Interferometry (VLBI), the relative positions of the Earth, Sun, and the pulsar can be determined from the ephemeris. (1) From this, we can know ϕ_{RS} in principle. (2) We can directly measure the angle Ψ_R at the Earth between the solar direction and the pulsar direction. (3) More importantly, the direction of radiating pulses that reaches the receiver can be also determined in principle because the viewing angle of the pulsar seen by the receiver is known from the pulse profiles. The viewing angle is changing with time because of the Earth's motion around the Sun. By using the pulsar position and the pulse radiation direction, we can determine Ψ_S. Please see Figure 3 for this situation. We explain in more detail how Ψ_S at S can be measured by the observer at R. We consider a pulsar of which the spin axis is known from some astronomical observations. A point is that the spin axis of an isolated pulsar is constant with time. The pulse shape and profile depend on the viewing angle with respect to the spin axis of the pulsar. The Earth moves around the Sun, and hence, the observer sees the same pulsar with different viewing angles with time. Accordingly, the observed pulse shape changes. By observing such a change in the pulse shape, we can in principle determine the intrinsic direction of the radio emission, namely the angle between the spin axis and the direction of the emitted light to the observer. In addition, we can know the intrinsic position (including the radial direction from the lens) of such a known pulsar from the ephemeris. By using the intrinsic position (its radial direction) and emission direction at S, Ψ_S can be determined in principle, though it is very difficult with current technology. As a result, we can determine in principle $\Psi_R - \Psi_S + \phi_{RS}$ from astronomical observations. Namely, α in Equation (7) is observable. Note that this procedure does not need to assume a different spacetime, while such a fiducial spacetime was assumed by Arakida (2018) [40], though the receiver in our universe cannot observe the fiducial different spacetime but can assume (or make theoretical calculations of) some quantities on the different spacetime.

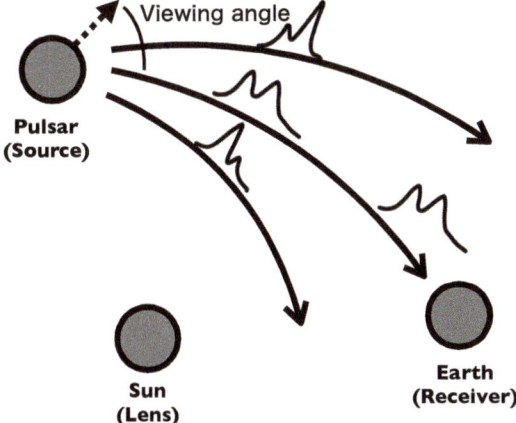

Figure 3. Observable α in Equation (7): In this schematic figure, the lens, receiver, and source are respectively the Sun, the Earth, and a pulsar that periodically radiates radio signals in a specific anistropic manner. From the pulse profile, we can determine the radiation direction at the source. By using the ephemeris, we know the relative positions of the Sun, Earth, and the pulsar. Hence, we can determine ϕ_{RS} and Ψ_S. By observing the pulsar, we can measure Ψ_R. In principle, therefore, we can determine $\Psi_R - \Psi_S + \phi_{RS}$ from these astronomical observations.

One can easily see that, in the far limit of the source and the receiver, Equation (9) agrees with the deflection angle of light as:

$$\alpha_\infty = 2 \int_0^{u_0} \frac{du}{\sqrt{F(u)}} - \pi. \tag{10}$$

Here, we define u and u_0 as as the inverse of r and the inverse of the closest approach (often denoted as r_0), respectively. $F(u)$ is defined as:

$$F(u) \equiv \left(\frac{du}{d\phi}\right)^2. \tag{11}$$

$F(u)$ can be computed by using Equation (3).

The present paper wishes to avoid the far limit for the following reason. Every observed star and galaxy is never located at infinite distance from us. For instance, we observe finite-redshift galaxies in cosmology. We cannot see objects at infinite redshift (exactly at the horizon). Except for a few rare cases in astronomy, the distance to the light source is much larger than the size of the lens. Therefore, we find a strong motivation for studying a situation in which the distance from the source to the receiver is finite. We define u_R and u_S as the inverse of r_R and r_S, respectively, where r_R and r_S are finite. Equation (7) is rewritten in an explicit form as [38,39]:

$$\alpha = \int_{u_R}^{u_0} \frac{du}{\sqrt{F(u)}} + \int_{u_S}^{u_0} \frac{du}{\sqrt{F(u)}} + \Psi_R - \Psi_S. \tag{12}$$

Here, we assume light rays that have only one local minimum of the radius coordinate between r_S and r_R. This is valid for normal situations in astronomy. However, we should note that multiple local minima are possible, e.g., if the emitter or the receiver (or both) are between the horizon and the light sphere in the Schwarzschild spacetime or if the emitter and receiver are at different sides of the throat of a wormhole spacetime. For such a case of multiple local minima, Equation (12) has to be modified because it assumes only the local minimum at $u = u_0$.

3. Weak Deflection of Light in Schwarzschild Spacetime

In this section, we consider the weak deflection of light in Schwarzschild spacetime, for which the line element becomes:

$$\begin{aligned} ds^2 &= -\left(1 - \frac{r_g}{r}\right) dt^2 + \frac{dr^2}{1 - \frac{r_g}{r}} \\ &\quad + r^2(d\theta^2 + \sin^2\theta d\phi^2), \end{aligned} \tag{13}$$

where $r_g = 2M$ in the geometrical unit. Then, $F(u)$ is:

$$F(u) = \frac{1}{b^2} - u^2 + r_g u^3. \tag{14}$$

By using Equation (6), $\Psi_R - \Psi_S$ in the Schwarzschild spacetime is expanded as:

$$\begin{aligned} \Psi_R^{Sch} - \Psi_S^{Sch} &\equiv [\arcsin(bu_R) + \arcsin(bu_S) - \pi] \\ &\quad -\frac{1}{2} br_g \left(\frac{u_R^2}{\sqrt{1 - b^2 u_R^2}} + \frac{u_S^2}{\sqrt{1 - b^2 u_S^2}}\right) + O(br_g^2 u_S^3, br_g^2 u_R^3). \end{aligned} \tag{15}$$

Note that $\Psi_R - \Psi_S \to \pi$ in the Schwarzschild spacetime as $u_S \to 0$ and $u_R \to 0$.

4. Other Examples

This section discusses two examples for a non-asymptotically flat spacetime. One is the Kottler solution to the Einstein Equation. The other is an exact solution in the Weyl conformal gravity. The aim of this study is to give us a suggestion or a speculation. We note that the present formulation is limited within the asymptotic flatness, rigorously speaking. As mentioned in the Introduction, Arakida [40] made an attempt to apply the Gauss–Bonnet theorem to quadrilaterals that are not extending to infinity, though a comparison between two different manifolds that he proposed is an open issue. A more careful study that gives a justification for this speculation or perhaps disproves it will be left for the future.

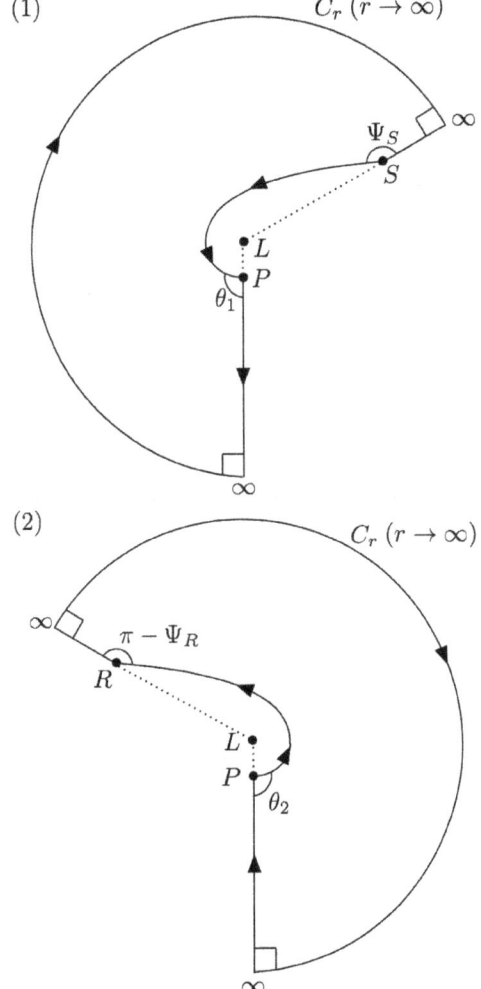

Figure 5. Quadrilaterals: They are made from the photon orbit in a non-Euclidean space. See Figure 4.

Next, we investigate a case of two loops shown by Figure 6. For this case, we add lines in order to divide the shape into four quadrilaterals as shown by Figure 7. We immediately find:

$$\begin{aligned}
\alpha^{(1)} &= (\pi - \theta_1) - \Psi_S + \phi_{RS}^{(1)}, \\
\alpha^{(2)} &= (\pi - \theta_3) - \theta_2 + \phi_{RS}^{(2)}, \\
\alpha^{(3)} &= (\pi - \theta_5) - \theta_4 + \phi_{RS}^{(3)}, \\
\alpha^{(4)} &= \Psi_R - \theta_6 + \phi_{RS}^{(4)},
\end{aligned} \quad (30)$$

where $\phi_{RS}^{(1)} + \phi_{RS}^{(2)} + \phi_{RS}^{(3)} + \phi_{RS}^{(4)} = \phi_{RS}$. Hence, we obtain:

$$\begin{aligned}
\alpha &= \alpha^{(1)} + \alpha^{(2)} + \alpha^{(3)} + \alpha^{(4)} \\
&= \Psi_R - \Psi_S + \phi_{RS},
\end{aligned} \quad (31)$$

where we use $\theta_1 + \theta_2 = \theta_3 + \theta_4 = \theta_5 + \theta_6 = \pi$. Equation (31) is obtained for the two-loop case in the same form as Equation (7). A loop does make contributions to α only through the terms of $\phi_{RS}^{(2)} + \phi_{RS}^{(3)}$.

Finally, we shall complete the proof. We consider the arbitrary winding number, say W. For this case, we prepare $2W$ quadrilaterals. We denote the inner angles at finite distance from L as $\theta_0, \cdots, \theta_{2W}$ in order from S to R as shown by Figure 8. Here, $\theta_0 = \Psi_S$ and $\theta_{2W} = \pi - \Psi_R$. Neighboring quadrilaterals (N) and (N+1) make the contribution to α only through $\phi_{RS}^{(N)} + \phi_{RS}^{(N+1)}$. We can understand this by noting that $\theta_{2N-1} + \theta_{2N} = \theta_{2N+1} + \theta_{2N+2} = \pi$, and the auxiliary lines cancel out. By induction, therefore, we complete the proof; Equation (7) holds for any winding number.

Equation (7) is equivalent to Equation (12). This is shown by using the orbit Equation. This expression is rearranged as:

$$\alpha = \Psi_R - \Psi_S + \phi_{RS}$$
$$= \Psi_R - \Psi_S + \int_{u_R}^{0} \frac{du}{\sqrt{F(u)}} + \int_{u_S}^{0} \frac{du}{\sqrt{F(u)}} + 2\int_{0}^{u_0} \frac{du}{\sqrt{F(u)}}. \quad (32)$$

We define the difference between the asymptotic deflection angle and the deflection angle for the finite distance case as $\delta\alpha$:

$$\delta\alpha \equiv \alpha - \alpha_\infty. \quad (33)$$

The meaning of this is the finite-distance correction to the deflection angle of light. By substituting Equations (10) and (32) into Equation (33), we get:

$$\delta\alpha = (\Psi_R - \Psi_S + \pi) + \int_{u_R}^{0} \frac{du}{\sqrt{F(u)}} + \int_{u_S}^{0} \frac{du}{\sqrt{F(u)}}. \quad (34)$$

This expression implies two origins of the finite-distance corrections. One origin is Ψ_R and Ψ_S. They are angles that are defined in a curved space. The other origin is the two path integrals. They contain information on the curved space. If we consider a receiver and source in the weak gravitational field (as common in astronomy), the finite-distance correction reflects only the weak field region, even if the light ray passes through a strong field region.

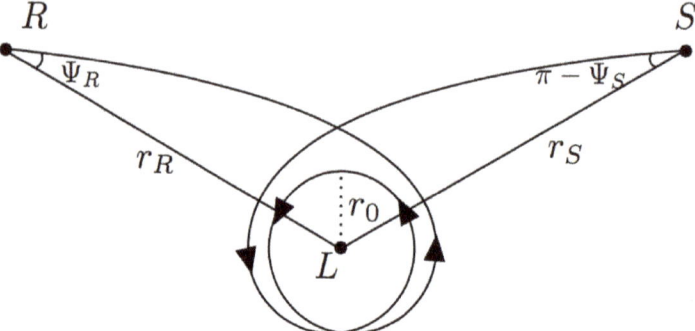

Figure 6. Two loops for the light ray in M^{opt}.

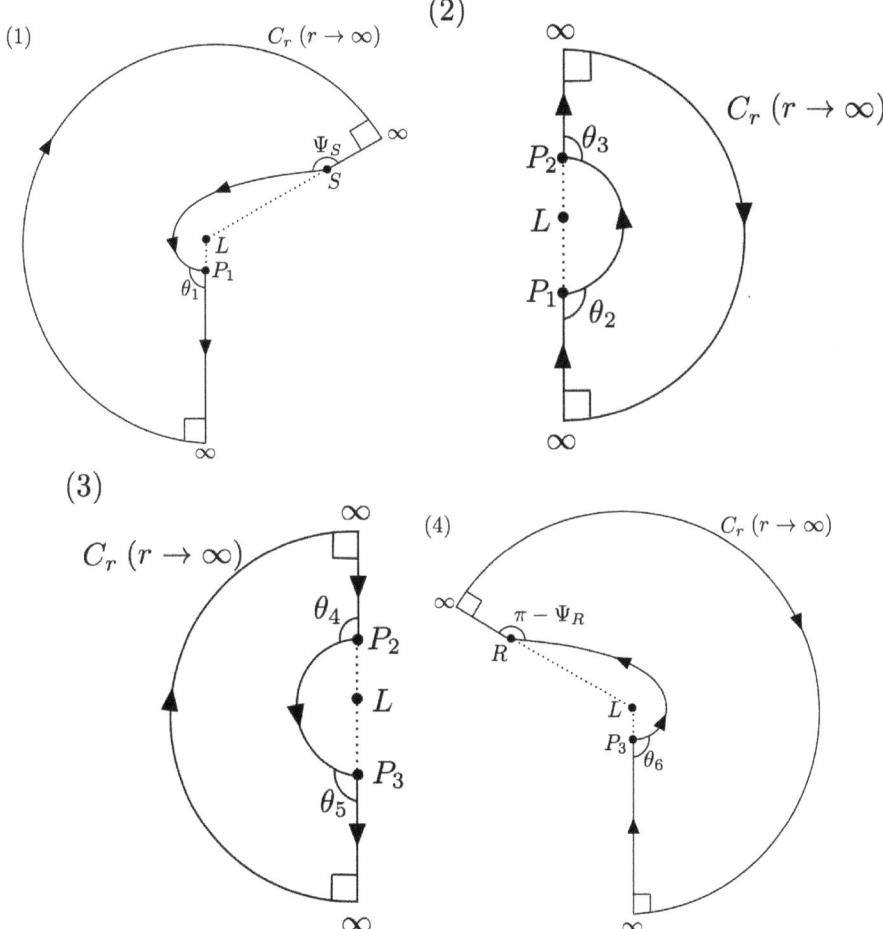

Figure 7. Quadrilaterals (1)–(4): They are in a non-Euclidean plane M^{opt}. See also Figure 6.

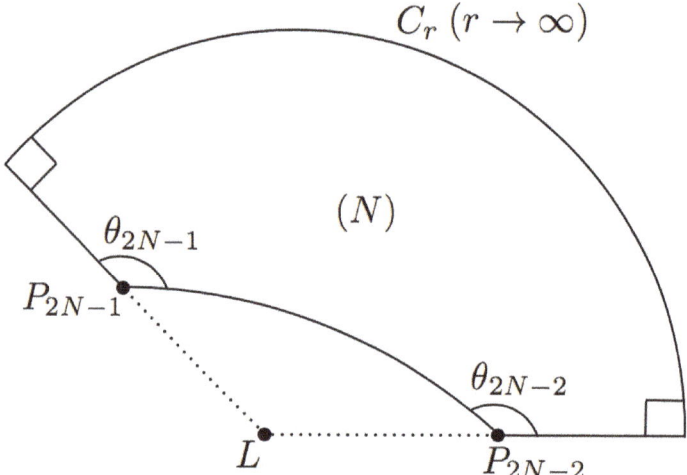

Figure 8. A quadrilateral in any loop number: This case is discussed when we prove by induction that Equation (7) holds in any loop number.

6. Strong Deflection of Light in Schwarzschild Spacetime

In this section, we consider the Schwarzschild black hole. By using $F(u)$ given by Equation (14), we solve Equation (32) in an analytic manner. The exact expressions involve incomplete elliptic integrals of the first kind. When the distances from the lens to the source and the receiver are much larger than the impact parameter of light ($r_S \gg b$, $r_R \gg b$) but the light ray passes near the photon sphere ($r_0 \sim 3M$), Equation (32) becomes approximately:

$$\alpha = \frac{2M}{b}\left[\sqrt{1 - b^2 u_R^2} + \sqrt{1 - b^2 u_S^2} - 2\right]$$
$$+ 2\log\left(\frac{12(2-\sqrt{3})r_0}{r_0 - 3M}\right) - \pi$$
$$+ O\left(\frac{M^2}{r_R^2}, \frac{M^2}{r_S^2}, 1 - \frac{3M}{r_0}\right), \qquad (35)$$

where we used a logarithmic term [8] in the last term of Equation (32). Here, the dominant terms in Ψ_R and Ψ_S cancel the terms in the integrals. As a consequence, Ψ_R and Ψ_S do not appear in the approximate expression of Equation (35).

As mentioned above, it follows that the logarithmic term by the strong gravity is free from finite-distance corrections such as $\sqrt{1-(bu_S)^2}$. By chance, $\delta\alpha$ in the strong deflection limit (See Equation (32)) is apparently the same as that for the weak deflection case (See, e.g., Equation (29) in Reference [39]). Therefore, the finite-distance correction in the strong deflection limit is again:

$$\delta\alpha \sim O\left(\frac{Mb}{r_S^2} + \frac{Mb}{r_R^2}\right). \qquad (36)$$

This is the same expression as that for the weak field case (e.g., Reference [38]). Namely, the correction is linear in the impact parameter. The finite-distance correction in the weak deflection case (large b) is thus larger than that in the strong deflection one (small b), if the other parameters remain the same.

*Sagittarius A**

Next, we briefly mention an astronomical implication of the strong deflection. One of the most feasible candidates for the strong deflection is Sagittarius * (Sgr A*) that is located at our galactic center. In this case, the receiver distance is much larger than the impact parameter of light and a source star may live in the bulge of our Galaxy.

The apparent size of Sgr A* is expected to be nearly the same as that of the central massive object of M87. However, the finite-distance correction to Sgr A* becomes much larger than that to the M87 case because Sgr A* is much closer to us than M87.

For Sgr A*, Equation (36) is evaluated as:

$$\delta \alpha \sim \frac{Mb}{r_S{}^2}$$

$$\sim 10^{-5} \text{arcsec.} \times \left(\frac{M}{4 \times 10^6 M_\odot}\right) \left(\frac{b}{3M}\right) \left(\frac{0.1 \text{pc}}{r_S}\right)^2, \tag{37}$$

where the central black hole mass is assumed as $M \sim 4 \times 10^6 M_\odot$ and we take the limit of strong deflection $b \sim 3M$. Rather interestingly, this correction as $\sim 10^{-5}$ arcsec. will be reachable by the Event Horizon Telescope [31–36] and near-future astronomy.

See Figure 9 for numerical estimations of the finite-distance correction by the source distance. This figure and Equation (37) suggest that $\delta \alpha$ is $\sim$ ten (or more) micro arcseconds if a source star is sufficiently close to Sgr A* for instance within a tenth of one parsec from Sgr A*. For such a case, the infinite-distance limit does not hold even though the source is still in the weak field. We should take account of finite-distance corrections that are discussed in this paper.

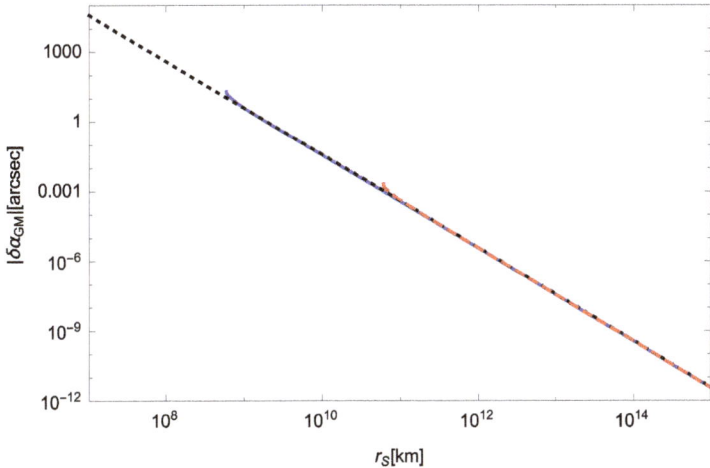

Figure 9. The finite-distance correction for Sgr A* as $\delta \alpha_{GM}$: The horizontal axis denotes the source distance r_S. The vertical one means the finite-distance correction to the light deflection. The solid line (blue in color) and dashed one (red in color) mean $b = 10^2 M$ and $b = 10^4 M$, respectively. The dotted curve (yellow in color) denotes the leading term of $\delta \alpha_{GM}$ given by Equation (33). These three lines are substantially overlapped with each other. This implies that $\delta \alpha_{GM}$ is weakly dependent on the impact parameter b.

In the strong deflection case, each orbit around the black hole will have a slightly different r_0, thereby producing a number of "ghost" images (often called relativistic images). In this paper, detailed calculations about it for the finite-distance source and receiver are not done. It is left for the future.

7. Defining the Gravitational Deflection Angle of Light for a Stationary and Axially Symmetric Spacetime

7.1. Optical Metric for the Stationary, Axisymmetric Spacetime

In this section, a stationary and axisymmetric spacetime is considered, for which we shall discuss how to define the gravitational deflection angle of light especially by using the Gauss–Bonnet theorem [42]. The line element in this spacetime is [58–60]:

$$\begin{aligned}ds^2 &= g_{\mu\nu}dx^\mu dx^\nu \\ &= -A(y^1,y^2)dt^2 - 2H(y^1,y^2)dt d\phi \\ &\quad + F(y^1,y^2)\gamma_{pq}dy^p dy^q + D(y^1,y^2)d\phi^2.\end{aligned} \quad (38)$$

Here, p and q mean 1 and 2, γ_{pq} is a two-dimensional symmetric tensor, μ, ν are from 0 to 3, and the t and ϕ coordinates respect the Killing vectors. We rewrite this metric into a form such that γ_{pq} can be diagonal. We prefer to use the polar coordinates rather than the cylindrical ones because the Kerr metric and the rotating Teo wormhole one are usually expressed in the polar coordinates. In this paper, we thus use the polar coordinates. In the cylindrical coordinates, the line element is known as the Weyl-Lewis-Papapetrou form [58–60]. Equation (38) is rewritten as

$$\begin{aligned}ds^2 &= -A(r,\theta)dt^2 - 2H(r,\theta)dt d\phi \\ &\quad + B(r,\theta)dr^2 + C(r,\theta)d\theta^2 + D(r,\theta)d\phi^2,\end{aligned} \quad (39)$$

where a local reflection symmetry is assumed with respect to the equatorial plane $\theta = \frac{\pi}{2}$.

This assumption is expressed as:

$$\left.\frac{\partial g_{\mu\nu}}{\partial \theta}\right|_{\theta=\frac{\pi}{2}} = 0. \quad (40)$$

The functions are $A(r,\theta) > 0, B(r,\theta) > 0, C(r,\theta) > 0, D(r,\theta) > 0$, and $H(r,\theta) > 0$. This assumption by Equation (40) is needed for the existence of a photon orbit on the equatorial plane. Note that we do not assume the global reflection symmetry with respect to the equatorial plane.

The null condition $ds^2 = 0$ is solved for dt as [61,62]:

$$dt = \sqrt{\gamma_{ij}dx^i dx^j} + \beta_i dx^i, \quad (41)$$

where i and j denote from 1 to 3 and γ_{ij} and β_i are defined as:

$$\gamma_{ij}dx^i dx^j \equiv \frac{B(r,\theta)}{A(r,\theta)}dr^2 + \frac{C(r,\theta)}{A(r,\theta)}d\theta^2 + \frac{A(r,\theta)D(r,\theta) + H^2(r,\theta)}{A^2(r,\theta)}d\phi^2, \quad (42)$$

$$\beta_i dx^i \equiv -\frac{H(r,\theta)}{A(r,\theta)}d\phi. \quad (43)$$

This spatial metric $\gamma_{ij}(\neq g_{ij})$ is used in order to define the arc length (ℓ) along the photon orbit as:

$$d\ell^2 \equiv \gamma_{ij}dx^i dx^j, \quad (44)$$

for which we define γ^{ij} by $\gamma^{ij}\gamma_{jk} = \delta^i{}_k$. γ_{ij} defines a 3-dimensional Riemannian space $^{(3)}M$, where the photon orbit is a spatial curve. In the appendix of Reference [62], they show that ℓ is an affine parameter of a light ray.

If the spacetime is static, spherically symmetric, and asymptotically flat, β_i is zero and γ_{ij} is nothing but the optical metric. The photon orbit follows a geodesic in a 3-dimensional Riemannian space. In this section and after, we refer to γ_{ij} as the generalized optical metric. Note that the metric γ_{ij} has been called the Fermat metric and the one-form β_i is called the Fermat one-form by some authors.

We apply the Gauss–Bonnet theorem to a surface (See Figure 1). The Gauss-Bonnet theorem is expressed as:

$$\iint_{\substack{R_\infty S_\infty \\ R \quad S}} K dS + \int_R^S \kappa_g d\ell + \int_{S_\infty}^{R_\infty} \bar{\kappa}_g d\ell + [\Psi_R + (\pi - \Psi_S) + \pi] = 2\pi, \tag{45}$$

where we note that the geodesic curvatures of the path from S to S_∞ and the path from R to R_∞ are both 0 because these paths are geodesic. κ_g is the geodesic curvature of the photon orbit, and $\bar{\kappa}_g$ is the geodesic curvature of the circular arc segment with an infinite radius.

7.2. Gaussian Curvature

In this subsection, we examine whether the rotational part (β_i) of the spacetime makes a contribution to the Gaussian curvature. The Gaussian curvature on the equatorial plane is expressed by using the 2-dimensional Riemann tensor $^{(2)}R_{r\phi r\phi}$:

$$\begin{aligned} K &= \frac{{}^{(2)}R_{r\phi r\phi}}{\det \gamma_{ij}^{(2)}} \\ &= \frac{1}{\sqrt{\det \gamma_{ij}^{(2)}}} \left[\frac{\partial}{\partial \phi} \left(\frac{\sqrt{\det \gamma_{ij}^{(2)}}}{\gamma_{rr}^{(2)}} {}^{(2)}\Gamma^\phi{}_{rr} \right) - \frac{\partial}{\partial r} \left(\frac{\sqrt{\det \gamma_{ij}^{(2)}}}{\gamma_{rr}^{(2)}} {}^{(2)}\Gamma^\phi{}_{r\phi} \right) \right], \end{aligned} \tag{46}$$

where $^{(2)}R_{r\phi r\phi}$ and $^{(2)}\Gamma^\beta{}_{jk}$ are defined by using the generalized optical metric γ_{ij} on the equatorial plane. $\det \gamma_{ij}^{(2)}$ is the determinant of the generalized optical metric in the equatorial plane.

dS in Equation (45) becomes:

$$dS = \sqrt{\det \gamma^{(2)}} dr d\phi. \tag{47}$$

The surface integration of the Gaussian curvature in Equation (45) is rewritten explicitly as:

$$\iint_{\substack{R_\infty \square S_\infty \\ R \quad S}} K dS = \int_{\phi_S}^{\phi_R} \int_{r_{OE}}^{\infty} K \sqrt{\det \gamma^{(2)}} dr d\phi, \tag{48}$$

where r_{OE} means the solution of the orbit Equation.

7.3. Geodesic Curvature

Let us imagine a parameterized curve in a surface. Roughly speaking, the geodesic curvature of the parameterized curve is a measure of how different the curve is from the geodesic. The geodesic curvature of the parameterized curve is defined as the surface-tangential component of the acceleration (namely the geodesic curvature) of the curve. The normal curvature is defined as the surface-normal component of the acceleration. The normal curvature does not appear in the present paper because we consider only the curves on the equatorial plane.

The geodesic curvature in the vector form is defined as (see, e.g., Reference [63,64]):

$$\kappa_g \equiv T' \cdot (T \times N), \tag{49}$$

where, for a parameterized curve, T denotes the unit tangent vector for the curve by reparameterizing the curve using its arc length, T' means its derivative with respect to the parameter, and N indicates the unit normal vector for the surface. The geodesic curvature of a curve vanishes if the curve follows the geodesic. This zero is because the acceleration vector T' vanishes.

7.4. Photon Orbit with the Generalized Optical Metric

In this subsection, we discuss geometrical aspects of a photon orbit in terms of the generalized optical metric. The unit vector tangent to the spatial curve is generally expressed as:

$$e^i \equiv \frac{dx^i}{d\ell}, \qquad (50)$$

where a parameter ℓ is defined by Equation (44).

The flight time T of a light from the source to the receiver is obtained by performing the integral of Equation (41):

$$T = \int_{t_S}^{t_R} dt = \int_S^R \left(\sqrt{\gamma_{ij} de^i de^j} + \beta_i de^i \right) d\ell. \qquad (51)$$

The light ray follows the Fermat's principle, namely $\delta T = 0$ [65]. The Lagrangian for a photon can be expressed as:

$$\mathcal{L} = \sqrt{\gamma_{ij} e^i e^j} + \beta_i e^i. \qquad (52)$$

From this, we obtain:

$$\frac{d}{d\ell} \frac{\partial \mathcal{L}}{\partial e^k} = \gamma_{ik} e^i_{,l} e^l + \gamma_{ik,l} e^i e^l + \beta_{k,i} e^i, \qquad (53)$$

$$\frac{\partial \mathcal{L}}{\partial x^k} = \frac{1}{2} \gamma_{ij,k} e^i e^j + \beta_{i,k} e^i, \qquad (54)$$

where we used $\gamma_{ij} e^i e^j = 1$ and the comma (,) defines the partial derivative. The Euler–Lagrange Equation is calculated as:

$$e^j_{,l} e^l + \gamma^{kj} \left(\gamma_{ik,l} e^i e^l - \frac{1}{2} \gamma_{il,k} e^i e^l \right) = \gamma^{kj} (\beta_{l,k} - \beta_{k,l}) e^l. \qquad (55)$$

This leads to the Equation for the light ray [62]:

$$\frac{de^i}{d\ell} = -\gamma^{il} (\gamma_{lj,k} - \frac{1}{2} \gamma_{jk,l}) e^j e^k + \gamma^{ij} (\beta_{k,j} - \beta_{j,k}) e^k.$$

Therefore, the geodesic Equation is equivalent to:

$$\begin{aligned} e^i_{|j} e^j &= \frac{de^i}{d\ell} + {}^{(3)}\Gamma^i{}_{jk} e^j e^k \\ &= \frac{de^i}{d\ell} + \gamma^{il} (\gamma_{lj,k} - \frac{1}{2} \gamma_{jk,l}) e^j e^k \\ &= \gamma^{ij} (\beta_{k,j} - \beta_{j,k}) e^k, \end{aligned} \qquad (56)$$

where we define $|$ as the covariant derivative with respect to γ_{ij}. ${}^{(3)}\Gamma^i{}_{jk}$ means the Christoffel symbol by γ_{ij}.

The acceleration vector a^i is defined by:

$$a^i \equiv e^i{}_{|j} e^j = \gamma^{ij} (\beta_{k|j} - \beta_{j|k}) e^k = \gamma^{ij} (\beta_{k,j} - \beta_{j,k}) e^k. \qquad (57)$$

By using the Levi–Civita symbol ε_{ijk}, we express the cross (outer) product: of A and B in the covariant manner:

$$\sqrt{\gamma}\varepsilon_{ijk}A^j B^k = (A \times B)_i. \tag{58}$$

The Levi–Civita tensor ϵ_{ijk} is defined by $\epsilon_{ijk} \equiv \sqrt{\gamma}\varepsilon_{ijk}$, where and ε_{ijk} is the Levi–Civita symbol ($\varepsilon_{123} = 1$).

The Levi–Civita tensor ϵ_{ijk} in a three-dimensional satisfies:

$$\epsilon_{sjk}\epsilon^{slm} = \sqrt{\gamma}\varepsilon_{sjk}\frac{1}{\sqrt{\gamma}}\varepsilon^{slm} = \delta_j^l \delta_k^m - \delta_j^m \delta_k^l, \tag{59}$$

$$\epsilon_{sjk}\epsilon^s{}_{lm} = \gamma_{jl}\gamma_{km} - \gamma_{jm}\gamma_{kl}. \tag{60}$$

By using Equations (58)–(60), Equation (57) is rewritten as:

$$a^i = \gamma^{ij}e^k \epsilon_{sjk}(\nabla \times \beta)^s. \tag{61}$$

Vector a^i is the spatial vector representing the acceleration due to β_i. In particular, a^i is caused in gravitomagnetism [66]. To be more precise, the gravitomagnetic vector has an analogy to the Lorentz force in electromagnetism $\propto v \times (\nabla \times A_m)$, in which A_m denotes the vector potential. The vector potential is defined as $B = \nabla \times A_m, E = -\nabla \phi - \frac{\partial A_m}{\partial t}$, where E and B are the electric and magnetic fields, respectively, and the electric potential is ϕ.

γ_{ij} is not an induced metric but the generalized optical metric. If β_i is nonvanishing, the photon orbit may be different from a geodesic in $^{(3)}M$ with γ_{ij}, even though the light ray in the four-dimensional spacetime follows the null geodesic.

In a stationary and axisymmetric spacetime, it is always possible to find out coordinates such that g_{0i} can vanish and $a^i = 0$. In this case, the photon orbit is considered a spatial geodesic curve in $^{(3)}M$.

We study axisymmetric cases which allow $g_{0i} \neq 0$. Therefore, geodesic curvature κ_g does not always vanish in the photon orbit in the Gauss–Bonnet theorem because the geodesic curvature κ_g for a photon orbit is due to the gravitomagnetic effect. This nonvanishing κ_g for the photon orbit leads to a crucial difference from the SSS case [38,39].

7.5. Geodesic Curvature of a Photon Orbit

Equation (49) is rearranged to be in the tensor form:

$$\kappa_g = \epsilon_{ijk} N^i a^j e^k, \tag{62}$$

where $\vec{T}$ and $\vec{T}'$ correspond to e^k and a^j, respectively.

In this paper, the acceleration vector of the photon orbit depends on β_i. Hence, the geodesic curvature for the photon orbit also depends on it. A nonvanishing integral of the geodesic curvature along the light ray appears in the Gauss–Bonnet theorem in Equation (8).

Substituting Equation (57) into a^i in Equation (62) leads to:

$$\begin{aligned}\kappa_g &= \epsilon_{ijk} N^i \gamma^{jl}(\beta_{n|l} - \beta_{l|n})e^n e^k \\ &= \gamma^{ja} N^i e^k e^b \epsilon_{ijk}\epsilon_{sab}\epsilon^{sml}\beta_{l|m} \\ &= N_i e_k e^b (\delta^i{}_s \delta^k{}_b - \delta^i{}_b \delta^k{}_s)\epsilon^{sml}\beta_{l|m} \\ &= -\epsilon^{ijk} N_i \beta_{j|k},\end{aligned} \tag{63}$$

where we used $\gamma_{ij}e^i e^j = 1$ and $\gamma_{ij}e^i N^j = 0$. The unit vector normal to the equatorial plane is:

$$N_p = \frac{1}{\sqrt{\gamma^{\theta\theta}}} \delta^\theta_p, \qquad (64)$$

where the upward direction is chosen without loss of generality.

For the equatorial plane, we obtain:

$$\epsilon^{\theta p q} \beta_{q|p} = -\frac{1}{\sqrt{\gamma}} \beta_{\phi,r}, \qquad (65)$$

where we use $\epsilon^{\theta r \phi} = -1/\sqrt{\gamma}$ and $\beta_{r,\phi} = 0$ because of the axisymmetry.

By using Equations (64) and (65), κ_g in Equation (63) becomes:

$$\kappa_g = -\frac{1}{\sqrt{\gamma \gamma^{\theta\theta}}} \beta_{\phi,r}. \qquad (66)$$

By using Equation (44), the line element in the path integral is obtained as:

$$d\ell = \sqrt{\gamma_{rr} \left(\frac{dr}{d\phi}\right)^2 + \gamma_{\phi\phi}} d\phi, \qquad (67)$$

where $\theta = \pi/2$.

7.6. Geodesic Curvature of a Circular arc Segment

In a flat space, the geodesic curvature κ of the circular arc segment of radius R is obtained as:

$$\kappa = \frac{1}{R}. \qquad (68)$$

The geodesic curvature $\bar{\kappa}_g$ of a circular arc segment of radius $R_c = R_\infty$ is obtained as:

$$\bar{\kappa}_g = \frac{1}{R_c}, \qquad (69)$$

where the radius R_c is sufficiently larger than r_R and r_S and the circular arc segment is in the asymptotically flat region.

Equation (44) becomes $d\ell^2 = dr^2 + r^2(d\theta^2 + \sin^2\theta d\phi^2)$ because we assume an asymptotically flat spacetime. Hence, the line element in the path integral of $\bar{\kappa}_g$ is obtained as:

$$d\ell = R_c d\phi, \qquad (70)$$

where we choose $\theta = \pi/2$ and $r = R_c$ for the circular arc segment.

Therefore, the path integral of $\bar{\kappa}_g$ in Equation (45) is rewritten as:

$$\int_{S_\infty}^{R_\infty} \bar{\kappa}_g d\ell = \int_{\phi_S}^{\phi_R} d\phi = \phi_R - \phi_S = \phi_{RS}, \qquad (71)$$

where we denote the angular coordinate values of the receiver and the source as ϕ_R and ϕ_S, respectively.

7.7. Impact Parameter and Light Rays

By using Equation (39), we study the orbit Equation on the equatorial plane. The Lagrangian for a photon in the equatorial plane is obtained as:

$$\hat{\mathcal{L}} = -A(r)\dot{t}^2 - 2H(r)\dot{t}\dot{\phi} + B(r)\dot{r}^2 + D(r)\dot{\phi}^2, \tag{72}$$

where the dot denotes the derivative with respect to the affine parameter and the functions $A(r), B(r), D(r)$, and $H(r)$ mean, to be rigorous, $A(r, \pi/2), B(r, \pi/2), D(r, \pi/2)$, and $H(r, \pi/2)$ respectively.

The metric (or the Lagrangian $\hat{\mathcal{L}}$ in the 4-dimensional spacetime) is independent from t and ϕ. Therefore:

$$\frac{d}{d\ell}\frac{\partial \hat{\mathcal{L}}}{\partial \dot{t}} = 0,$$

$$\frac{d}{d\ell}\frac{\partial \hat{\mathcal{L}}}{\partial \dot{\phi}} = 0.$$

Then, associated with the two Killing vectors $\xi^\mu = (1,0,0,0)$ and $\bar{\xi}^\mu = (0,0,0,1)$, respectively:

$$\frac{\partial \hat{\mathcal{L}}}{\partial \dot{t}} = g_{\mu\nu}\xi^\mu k^\nu,$$

$$\frac{\partial \hat{\mathcal{L}}}{\partial \dot{\phi}} = g_{\mu\nu}\bar{\xi}^\mu k^\nu, \tag{73}$$

where $k^\mu = \frac{dx^\mu}{d\ell}$ is the vector tangent to the light ray in the four-dimensional spacetime. There are two constants of motion:

$$E = A(r)\dot{t} + H(r)\dot{\phi}, \tag{74}$$
$$L = D(r)\dot{\phi} - H(r)\dot{t}, \tag{75}$$

where E denotes the energy of the photon and L means the angular momentum of the photon. The impact parameter of the photon is defined as:

$$b \equiv \frac{L}{E}$$
$$= \frac{-H(r)\dot{t} + D(r)\dot{\phi}}{A(r)\dot{t} + H(r)\dot{\phi}}$$
$$= \frac{-H(r) + D(r)\frac{d\phi}{dt}}{A(r) + H(r)\frac{d\phi}{dt}}. \tag{76}$$

In terms of the impact parameter b, $\hat{\mathcal{L}} = 0$ can be considered as the orbit Equation:

$$\left(\frac{dr}{d\phi}\right)^2 = \frac{A(r)D(r) + H^2(r)}{B(r)} \frac{D(r) - 2H(r)b - A(r)b^2}{[H(r) + A(r)b]^2}, \tag{77}$$

where we used Equation (39). By introducing $u \equiv 1/r$, we rewrite the orbit Equation as:

$$\left(\frac{du}{d\phi}\right)^2 = F(u), \tag{78}$$

where $F(u)$ is:

$$F(u) = \frac{u^4(AD+H^2)(D-2Hb-Ab^2)}{B(H+Ab)^2}. \tag{79}$$

We examine the angles (Ψ_R and Ψ_S in Figure 10) at the receiver position and the source one. The unit vector tangent to the photon orbit in $^{(3)}M$ is e^i. Its components on the equatorial plane are expressed as:

$$e^i = \frac{1}{\zeta}\left(\frac{dr}{d\phi}, 0, 1\right), \tag{80}$$

where ζ satisfies:

$$\frac{1}{\zeta} = \frac{A(r)[H(r)+A(r)b]}{A(r)D(r)+H^2(r)}. \tag{81}$$

This can be derived also from $\gamma_{ij}e^i e^j = 1$ by using Equation (77).

In the equatorial plane, the unit radial vector is:

$$R^i = \left(\frac{1}{\sqrt{\gamma_{rr}}}, 0, 0\right), \tag{82}$$

where the outgoing direction is chosen for a sign convention.

By using the inner product between e^i and R^i, we therefore define the angle as:

$$\cos\Psi \equiv \gamma_{ij}e^i R^j$$
$$= \sqrt{\gamma_{rr}}\frac{A(r)[H(r)+A(r)b]}{A(r)D(r)+H^2(r)}\frac{dr}{d\phi}, \tag{83}$$

where Equations (80)–(82) are used. This is rewritten as:

$$\sin\Psi = \frac{H(r)+A(r)b}{\sqrt{A(r)D(r)+H^2(r)}}, \tag{84}$$

where Equation (77) is used. We should note that $\sin\Psi$ in Equation (84) is more useful in practical calculations, because it needs only the local quantities. On the other hand, $\cos\Psi$ by Equation (83) needs the derivative $dr/d\phi$. In addition, the domain of this Ψ is $0 \leq \Psi \leq \pi$ and, hence, $\sin\Psi$ is always positive.

By substituting r_R and r_S into r of Equation (84), we obtain $\sin\Psi_R$ and $\sin\Psi_S$, respectively. We note that the range of the principal value of $y = \arcsin x$ is $-\frac{\pi}{2} \leq y \leq \frac{\pi}{2}$ as usual. However, the range of Ψ_R (Ψ_S) is $0 \leq \Psi_R(\Psi_S) \leq \pi$. By using the usual principal value, Equation (84) for (Ψ_R) and (Ψ_S) becomes:

$$\sin\Psi_R = \frac{H(r_R)+A(r_R)b}{\sqrt{A(r_R)D(r_R)+H^2(r_R)}}, \tag{85}$$

$$\sin(\pi-\Psi_S) = \frac{H(r_S)+A(r_S)b}{\sqrt{A(r_S)D(r_S)+H^2(r_S)}}, \tag{86}$$

respectively because Ψ_R is an acute angle and Ψ_S is an obtuse angle as shown by Figure 10.

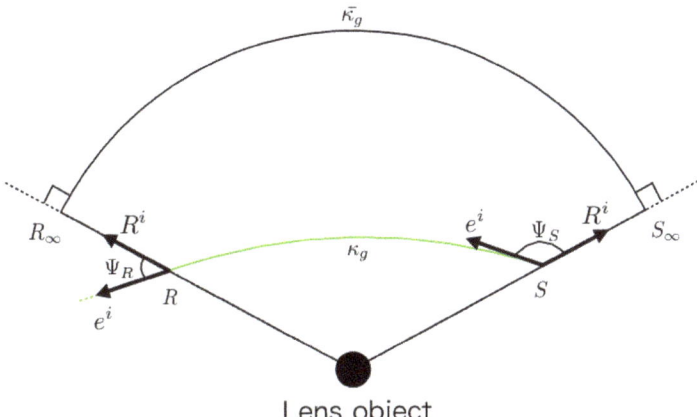

Figure 10. Ψ_R and Ψ_S: Ψ_R is the angle between the radial direction and the light ray at the receiver position. Ψ_S is that at the source position.

7.8. Gravitational Deflection Light in the Axisymmetric Case

We define:
$$\alpha \equiv \Psi_R - \Psi_S + \phi_{RS} \tag{87}$$

for the equatorial plane in the axisymmetric spacetime. This definition apparently depends on the angular coordinate ϕ. By using the Gauss–Bonnet theorem in Equation (8), this Equation is rearranged as:

$$\alpha = -\iint_{{}^\infty_R \square {}^\infty_S} K dS - \int_R^S \kappa_g d\ell. \tag{88}$$

Here, $d\ell$ is positive when the photon is in the prograde motion, whereas it is negative for the retrograde case. Equation (88) means that α is coordinate-invariant for the axisymmetric case. Up until now, we did not use any Equations for gravitational fields. Therefore, the above discussion and results still stand not only in the theory of general relativity but also in a general class of metric theories of gravity only if the light ray in the four-dimensional spacetime is a null geodesic.

8. Weak Deflection of Light in Kerr Spacetime

8.1. Kerr Spacetime and γ_{ij}

In this section, we focus on the weak deflection of light in the Kerr spacetime as an axisymmetric example. Kerr metric in the Boyer–Lindquist form is expressed as:

$$ds^2 = -\left(1 - \frac{2Mr}{\Sigma}\right)dt^2 - \frac{4aMr\sin^2\theta}{\Sigma}dtd\phi \\ + \frac{\Sigma}{\Delta}dr^2 + \Sigma d\theta^2 + \left(r^2 + a^2 + \frac{2a^2Mr\sin^2\theta}{\Sigma}\right)\sin^2\theta d\phi^2, \tag{89}$$

where Σ and Δ are defined as:

$$\Sigma \equiv r^2 + a^2 \cos^2 \theta, \tag{90}$$

$$\Delta \equiv r^2 - 2Mr + a^2. \tag{91}$$

Using the Gauss–Bonnet theorem, the deflection angle of light in the Kerr spacetime was calculated for the asymptotic source and receiver by Werner [67]. However, his method based on the osculating metric is limited within the asymptotic case. Later, Ono et al. developed a different approach using the Gauss–Bonnet theorem that enables the calculation of the deflection angle for the finite distance case in the Kerr spacetime [42].

By using Equations (42) and (43), the generalized optical metric γ_{ij} and the gravitomagnetic term β_i for the Kerr metric are obtained as:

$$\gamma_{ij} dx^i dx^j = \frac{\Sigma^2}{\Delta(\Sigma - 2Mr)} dr^2 + \frac{\Sigma^2}{(\Sigma - 2Mr)} d\theta^2 \\ + \left(r^2 + a^2 + \frac{2a^2 Mr \sin^2 \theta}{(\Sigma - 2Mr)} \right) \frac{\Sigma \sin^2 \theta}{(\Sigma - 2Mr)} d\phi^2, \tag{92}$$

$$\beta_i dx^i = -\frac{2aMr \sin^2 \theta}{(\Sigma - 2Mr)} d\phi. \tag{93}$$

Note that γ_{ij} has no linear terms in the Kerr spin parameter a because only g_{0i} in $g_{\mu\nu}$ has a linear term in a and $g_{0i} \propto H$ contributes to γ_{ij} through a quadratic term $g_{0i} g_{0j} \propto H^2$, as shown by Equation (42).

In order to calculate the Gaussian curvature K of the equatorial plane, the geodesic curvature κ_g of the light ray and the geodesic curvature $\bar{\kappa}_g$ of the circular arc of an infinite radius and of the angles Ψ_R and Ψ_S, we use two approximations for the weak field and slow rotation, where M and a play roles as book-keeping parameters though they are dimensional quantities.

By using Equation (77), we obtain the orbit Equation:

$$\left(\frac{dr}{d\phi}\right)^2 = \frac{b^2 \left\{ \frac{a^2}{b^2} + \frac{r}{b}\left(\frac{r}{b} - \frac{2M}{b}\right) \right\}^2 \left\{ \frac{a^2}{b^2}\left(\frac{2M}{b} + \frac{r}{b}\right) - \frac{4aM}{b^2} + \frac{2M}{b} - \frac{r}{b} + \frac{r^3}{b^3} \right\}}{\frac{r}{b}\left\{ \frac{2aM}{b^2} + \frac{r}{b} - \frac{2M}{b} \right\}^2} \\ = \frac{r^4}{b^2} - r^2 + 2Mr - \frac{4r^3}{b^3} aM + \mathcal{O}(a^2), \tag{94}$$

where the weak-field and slow-rotation approximations are used in the last line. There are no M-squared terms in the last line. The orbit Equation becomes:

$$\left(\frac{du}{d\phi}\right)^2 = F(u) = \frac{1}{b^2} - u^2 + 2Mu^3 - \frac{4u}{b^3} aM + \mathcal{O}(a^2 u^4). \tag{95}$$

We solve iteratively Equation (95). In order to find the zeroth-order solution, we solve the truncated Equation (95):

$$\left(\frac{du}{d\phi}\right)^2 = \frac{1}{b^2} - u^2 + \mathcal{O}(Mu^3, aMu^4, a^2u^4). \tag{96}$$

The zeroth-order solution for this Equation is:

$$u = \frac{\sin \phi}{b}, \tag{97}$$

where we use $\left.\frac{du}{d\phi}\right|_{\phi=\pi/2} = 0$ as the boundary condition. This condition means that the closest approach of the photon orbit is expressed as $r = r_0 = 1/u_0, \phi = \pi/2$. We assume that the linear-order solution with M is $u = \frac{\sin\phi}{b} + u_1(\phi)M$. In order to obtain $u_1(\phi)$, we substitute this expression of u into the Equation (95) with terms linear in M:

$$\left(\frac{du}{d\phi}\right)^2 = \frac{1}{b^2} - u^2 + 2Mu^3 + \mathcal{O}(aMu^4, a^2u^4). \tag{98}$$

$u_1(\phi)$ is thus obtained as:

$$u_1(\phi) = \frac{1}{b^2}(1 + \cos^2\phi), \tag{99}$$

where we used the boundary condition mentioned above. The solution with a is in a form of $u = \frac{\sin\phi}{b} + \frac{M}{b^2}(1 + \cos^2\phi) + u_2(\phi)a$. Since Equation (95) does not include any linear term in a, we find $u_2(\phi) = 0$. The solution with aM is $u = \frac{\sin\phi}{b} + \frac{M}{b^2}(1 + \cos^2\phi) + u_3(\phi)aM$. We substitute this solution into Equation (95):

$$\frac{aM}{b}\left\{b^3\frac{du_3(\phi)}{d\phi}\cos\phi + b^3 u_3(\phi)\sin\phi + 2\sin\phi\right\} + \mathcal{O}(a^2u^4) = 0. \tag{100}$$

Hence, $u_3(\phi)$ is obtained as:

$$u_3(\phi) = -\frac{2}{b^3}. \tag{101}$$

Bringing the above results together, the iterative solution of Equation (95) is expressed as:

$$u = \frac{\sin\phi}{b} + \frac{M}{b^2}(1 + \cos^2\phi) - \frac{2aM}{b^3} + \mathcal{O}\left(\frac{M^2}{b^3}, \frac{a^2}{b^3}\right). \tag{102}$$

Next, we solve Equation (102) for ϕ. We obtain ϕ as:

$$\phi = \begin{cases} \arcsin(bu) + \frac{-2+b^2u^2}{b\sqrt{1-b^2u^2}}M + \frac{2aM}{b^2\sqrt{1-b^2u^2}} + \mathcal{O}\left(\frac{M^2}{b^3}, \frac{a^2}{b^3}\right) & (|\phi| < \frac{\pi}{2}) \\ \pi - \arcsin(bu) - \frac{-2+b^2u^2}{b\sqrt{1-b^2u^2}}M - \frac{2aM}{b^2\sqrt{1-b^2u^2}} + \mathcal{O}\left(\frac{M^2}{b^3}, \frac{a^2}{b^3}\right) & (\frac{\pi}{2} < |\phi|) \end{cases}, \tag{103}$$

where we can choose the domain of ϕ to be $-\pi \leq \phi < \pi$ without loss of generality. In the following, the range of the angular coordinate value ϕ_S at the source point is $-\frac{\pi}{2} \leq \phi_S < \frac{\pi}{2}$ and the range of the angular coordinate value ϕ_R at the receiver point is $|\phi_R| > \frac{\pi}{2}$. We find $|bu| < 1$ because the square root in Equation (103) must be real and nonzero, and the values of b and u are positive. Therefore, bu satisfies $0 < bu < 1$ in our calculation.

8.2. Gaussian Curvature on the Equatorial Plane

Let us explain how to compute the Gaussian curvature by using Equation (46). In the Kerr case, it becomes:

$$K = \frac{M\left(-6r\left(a^2 + M^2\right) + 6a^2M + 7Mr^2 - 2r^3\right)}{r^5(r - 2M)}$$
$$= -\frac{2M}{r^3} + \mathcal{O}\left(\frac{M^2}{r^4}, \frac{a^2M}{r^5}\right), \tag{104}$$

where the weak-field and slow-rotation approximations are used in the last line.

Next, we discuss the area element on the equatorial plane by using Equation (47). In the Kerr case, the area element of the equatorial plane is expressed as:

$$dS = [r + 3M + \mathcal{O}(M^2/r)]drd\phi. \tag{105}$$

By using Equations (104) and (105), the surface integral of the Gaussian curvature in Equation (88) is performed as:

$$-\iint_{{}_R^{R_\infty}\square_S^{S_\infty}} KdS = \int_{\phi_S}^{\phi_R} \int_{\infty}^{r_{OE}} (-\frac{2M}{r^3}r)drd\phi + \mathcal{O}\left(\frac{M^2}{b^2}, \frac{aM^2}{b^3}, \frac{a^2M}{b^3}\right)$$

$$= 2M \int_{\phi_S}^{\phi_R} \int_0^{\frac{1}{b}\sin\phi + \frac{M}{b^2}(1+\cos^2\phi) - \frac{2aM}{b^3}} dud\phi + \mathcal{O}\left(\frac{M^2}{b^2}, \frac{aM^2}{b^3}, \frac{a^2M}{b^3}\right)$$

$$= 2M \int_{\phi_S}^{\phi_R} \left[\frac{1}{b}\sin\phi\right]d\phi + \mathcal{O}\left(\frac{M^2}{b^2}, \frac{aM^2}{b^3}, \frac{a^2M}{b^3}\right)$$

$$= \frac{2M}{b}\left[\cos\phi_S - \cos\phi_R\right] + \mathcal{O}\left(\frac{M^2}{b^2}, \frac{aM^2}{b^3}, \frac{a^2M}{b^3}\right)$$

$$= \frac{2M}{b}\left[\sqrt{1-b^2u_S^2} + \sqrt{1-b^2u_R^2}\right] + \mathcal{O}\left(\frac{M^2}{b^2}, \frac{aM^2}{b^3}, \frac{a^2M}{b^3}\right), \tag{106}$$

where r_{OE} in the first line is the solution of Equation (94), we transform the integral variable as $r = 1/u$ in the second line, and we used $\cos\phi_S = \sqrt{1-b^2u_S^2} + \mathcal{O}(M/b)$ and $\cos\phi_R = -\sqrt{1-b^2u_R^2} + \mathcal{O}(M/b)$ from Equation (103) in the last line.

8.3. Path Integral of κ_g

Substituting Equation (93) into β_i in Equation (66) leads to:

$$\kappa_g = -\frac{2aM}{r^2(r-2M)}\left(\frac{1 - \frac{2M}{r} + \frac{a^2}{r^2}}{1 + \frac{a^2}{r^2} + \frac{2a^2M}{r^3}}\right)^{1/2}$$

$$= -\frac{2aM}{r^3} + \mathcal{O}\left(\frac{aM^2}{r^4}\right), \tag{107}$$

where the weak-field and slow-rotation approximations are used in the last line. We stress that the terms of a^nM ($n \geq 2$) do not exist in this expression.

The line element for the path integral by Equation (67) becomes:

$$d\ell = \left[\frac{b}{\sin^2\phi} + \mathcal{O}(M)\right]d\phi, \tag{108}$$

where Equation (102) was used for a relation between r and ϕ.

By using Equations (107) and (108), the path integral of κ_g in Equation (88) is performed as:

$$-\int_R^S \kappa_g d\ell = -\int_S^R \frac{2aM}{r^3}d\ell + \mathcal{O}\left(\frac{aM^2}{r^4}\right)$$

$$= -\frac{2aM}{b^2}\int_{\phi_S}^{\phi_R}\sin\phi d\phi + \mathcal{O}\left(\frac{aM^2}{r^4}\right)$$

$$= -\frac{2aM}{b^2}[\sqrt{1-b^2u_R^2} + \sqrt{1-b^2u_S^2}] + \mathcal{O}\left(\frac{aM^2}{b^3}\right). \tag{109}$$

Here, we assumed $d\ell > 0$, such that the orbital angular momentum can be parallel with the spin of the black hole, and we used a linear approximation of the photon orbit as $1/r = u = \sin\phi/b + \mathcal{O}(M/b^2, aM/b^3)$ from Equation (102). In the retrograde case, $d\ell$ becomes negative and the magnitude of the above path integral thus remains the same but the sign of the integral is opposite.

8.4. ϕ_{RS} Part

The displacement of the angular coordinate ϕ in Equation (87) is computed as:

$$\phi_{RS} = \int_S^R d\phi$$
$$= 2\int_0^{u_0} \frac{1}{\sqrt{F(u)}} du + \int_{u_S}^0 \frac{1}{\sqrt{F(u)}} du + \int_{u_R}^0 \frac{1}{\sqrt{F(u)}} du, \tag{110}$$

where the orbit equation of Equation (78) was used. We substitute Equation (95) into $F(u)$ in Equation (110) to obtain:

$$\phi_{RS} = \int_{u_S}^{u_0} \left(\frac{1}{\sqrt{u_0^2 - u^2}} + M \frac{u_0^3 - u^3}{(u_0^2 - u^2)^{3/2}} - 2aM \frac{u_0^3(u_0 - u)}{(u_0^2 - u^2)^{3/2}} \right) du$$
$$+ \int_{u_R}^{u_0} \left(\frac{1}{\sqrt{u_0^2 - u^2}} + M \frac{u_0^3 - u^3}{(u_0^2 - u^2)^{3/2}} - 2aM \frac{u_0^3(u_0 - u)}{(u_0^2 - u^2)^{3/2}} \right) du$$
$$+ \mathcal{O}(M^2 u_0^2, a^2 u_0^2)$$
$$= \left(\frac{\pi}{2} - \arcsin\left(\frac{u_S}{u_0}\right) + M \frac{(2u_0 + u_S)\sqrt{u_0^2 - u_S^2}}{u_0 + u_S} - 2aM \frac{u_0^3 \sqrt{u_0^2 - u_S^2}}{u_0^2 + u_0 u_S} \right)$$
$$+ \left(\frac{\pi}{2} - \arcsin\left(\frac{u_R}{u_0}\right) + M \frac{(2u_0 + u_R)\sqrt{u_0^2 - u_R^2}}{u_0 + u_R} - 2aM \frac{u_0^3 \sqrt{u_0^2 - u_R^2}}{u_0^2 + u_0 u_R} \right)$$
$$+ \mathcal{O}\left(M^2 u_0^2, a^2 u_0^2\right), \tag{111}$$

where the prograde case is assumed. In the retrograde motion, the sign of the linear term in a is opposite. In Equation (111), the impact parameter b is rewritten in terms of the closest approach u_0 for the integration from u_S (or u_R) to u_0. Namely, Equation (95) tells us the relation between the impact parameter b and the inverse of the closest approach u_0 as $b = u_0^{-1} + M - 2aMu_0 + \mathcal{O}(M^2 u_0, a^2 u_0)$ in the weak-field and slow-rotation approximations. By making use of this relation, Equation (111) is rearranged as:

$$\phi_{RS} = \pi - \arcsin(bu_S) - \arcsin(bu_R) + \frac{M(2 - b^2 u_S^2)}{b\sqrt{1 - b^2 u_S^2}} + \frac{M(2 - b^2 u_R^2)}{b\sqrt{1 - b^2 u_R^2}}$$
$$- \frac{2aM}{b^2} \left[\frac{1}{\sqrt{1 - b^2 u_S^2}} + \frac{1}{\sqrt{1 - b^2 u_R^2}} \right] + \mathcal{O}\left(M^2/b^2, a^2/b^2\right). \tag{112}$$

The first line of this equation recovers Equation (32) of Reference [38].

8.5. Ψ Parts

In the Kerr spacetime by Equation (89), Equation (85) is:

$$\sin \Psi_R = \frac{b}{r_R} \times \frac{1 - \frac{2M}{r_R} + \frac{2aM}{br_R}}{\sqrt{1 - \frac{2M}{r_R} + \frac{a^2}{r_R^2}}},$$

$$= \frac{b}{r_R}\left(1 - \frac{M}{r_R} + \frac{2aM}{br_R}\right) + \mathcal{O}\left(\frac{M^2}{r_R^2}, \frac{a^2}{r_R^2}, \frac{aM^2}{r_R^3}\right)$$

$$= bu_R\left(1 - Mu_R + \frac{2aMu_R}{b}\right) + \mathcal{O}\left(M^2 u_R^2, a^2 u_R^2, aM^2 u_R^3\right), \tag{113}$$

and Equation (86) is calculated as:

$$\sin(\pi - \Psi_S) = bu_S\left(1 - Mu_S + \frac{2aMu_S}{b}\right) + \mathcal{O}\left(M^2 u_S^2, a^2 u_S^2, aM^2 u_S^3\right), \tag{114}$$

where $r_R = 1/u_R$, $r_S = 1/u_S$ and we used the weak-field and slow-rotation approximations. By combining Equations (113) and (114), we obtain Ψ_R and Ψ_S as:

$$\Psi_R = \arcsin\left[bu_R\left(1 - Mu_R + \frac{2aMu_R}{b}\right)\right] + \mathcal{O}\left(M^2 u_R^2, a^2 u_R^2, aM^2 u_R^3\right)$$

$$= \arcsin(bu_R) - \frac{Mbu_R^2}{\sqrt{1 - b^2 u_R^2}} + \frac{2aMu_R^2}{\sqrt{1 - b^2 u_R^2}} + \mathcal{O}\left(M^2 u_R^2, a^2 u_R^2, aM^2 u_R^3\right),$$

$$\pi - \Psi_S = \arcsin(bu_S) - \frac{Mbu_S^2}{\sqrt{1 - b^2 u_S^2}} + \frac{2aMu_S^2}{\sqrt{1 - b^2 u_S^2}} + \mathcal{O}\left(M^2 u_S^2, a^2 u_S^2, aM^2 u_S^3\right). \tag{115}$$

By combining these relations, we obtain the Ψ part in Equation (87) as:

$$\Psi_R - \Psi_S = \arcsin(bu_R) + \arcsin(bu_S) - \pi - \frac{Mbu_R^2}{\sqrt{1 - b^2 u_R^2}} - \frac{Mbu_S^2}{\sqrt{1 - b^2 u_S^2}}$$

$$+ \frac{2aMu_R^2}{\sqrt{1 - b^2 u_R^2}} + \frac{2aMu_S^2}{\sqrt{1 - b^2 u_S^2}} + \mathcal{O}\left(M^2 u_R^2, M^2 u_S^2, a^2 u_R^2, a^2 u_S^2, aM^2 u_R^3, aM^2 u_S^3\right). \tag{116}$$

8.6. Deflection of Light in Kerr Spacetime

On the equatorial plane in the Kerr spacetime, the deflection angle of light is described by Equations (87) and (88). Let us examine whether the two results agree with each other.

First, we substitute Equations (112) and (116) into Equation (87). We obtain the deflection angle of light as:

$$\begin{aligned}\alpha_{prog} =& \arcsin(bu_R) + \arcsin(bu_S) - \pi - \frac{Mbu_R^2}{\sqrt{1-b^2u_R^2}} - \frac{Mbu_S^2}{\sqrt{1-b^2u_S^2}} \\ &+ \frac{2aMu_R^2}{\sqrt{1-b^2u_R^2}} + \frac{2aMu_S^2}{\sqrt{1-b^2u_S^2}} \\ &+ \pi - \arcsin(bu_S) - \arcsin(bu_R) + \frac{M(2-b^2u_S^2)}{b\sqrt{1-b^2u_S^2}} + \frac{M(2-b^2u_R^2)}{b\sqrt{1-b^2u_R^2}} \\ &- \frac{2aM}{b^2}\left[\frac{1}{\sqrt{1-b^2u_S^2}} + \frac{1}{\sqrt{1-b^2u_R^2}}\right] + \mathcal{O}\left(\frac{M^2}{b^2}\right) \\ =& \frac{2M}{b}\left(\sqrt{1-b^2u_R^2} + \sqrt{1-b^2u_S^2}\right) \\ &- \frac{2aM}{b^2}\left(\sqrt{1-b^2u_R^2} + \sqrt{1-b^2u_S^2}\right) + \mathcal{O}\left(\frac{M^2}{b^2}\right),\end{aligned} \quad (117)$$

where the prograde orbit of light is assumed. For the retrograde motion, we obtain:

$$\begin{aligned}\alpha_{retro} =& \frac{2M}{b}\left(\sqrt{1-b^2u_R^2} + \sqrt{1-b^2u_S^2}\right) \\ &+ \frac{2aM}{b^2}\left(\sqrt{1-b^2u_R^2} + \sqrt{1-b^2u_S^2}\right) + \mathcal{O}\left(\frac{M^2}{b^2}\right).\end{aligned} \quad (118)$$

Next, we substitute Equations (106) and (109) into Equation (88). Then, we obtain the deflection angle of light in the prograde motion as:

$$\begin{aligned}\alpha_{prog} =& \frac{2M}{b}\left(\sqrt{1-b^2u_R^2} + \sqrt{1-b^2u_S^2}\right) \\ &- \frac{2aM}{b^2}\left(\sqrt{1-b^2u_R^2} + \sqrt{1-b^2u_S^2}\right) + \mathcal{O}\left(\frac{M^2}{b^2}\right),\end{aligned} \quad (119)$$

and the deflection angle for the retrograde case as:

$$\begin{aligned}\alpha_{retro} =& \frac{2M}{b}\left(\sqrt{1-b^2u_R^2} + \sqrt{1-b^2u_S^2}\right) \\ &+ \frac{2aM}{b^2}\left(\sqrt{1-b^2u_R^2} + \sqrt{1-b^2u_S^2}\right) + \mathcal{O}\left(\frac{M^2}{b^2}\right).\end{aligned} \quad (120)$$

Note that the a^2 terms in the deflection angle in Equation (87) cancel out thanks to Equation (88). Here, we consider the limit as $u_R \to 0$ and $u_S \to 0$. In this limit, we get:

$$\alpha_{\infty\, prog} \to \frac{4M}{b} - \frac{4aM}{b^2} + \mathcal{O}\left(\frac{M^2}{b^2}\right), \quad (121)$$

$$\alpha_{\infty\, retro} \to \frac{4M}{b} + \frac{4aM}{b^2} + \mathcal{O}\left(\frac{M^2}{b^2}\right). \quad (122)$$

This shows that Equations (117) and (118) agree with the asymptotic deflection angles that are known in earlier works [4,68–70]. Precise analytic treatments of the deflection angle of light were done in a conventional approach, on the equatorial plane of a Kerr black hole [70] and for generic photon orbits in terms of the generalized hypergeometric functions of Appell and Lauricella [71]. They assume that both the source and the receiver are located at the null infinity.

If we wish to consider the deflection angle of light in a case where the receiver point is closer to the source point than the closest approach point, Equations (117) and (118) become:

$$\alpha_{prog} = \frac{2M}{b}\left(\sqrt{1-b^2 u_S^2} - \sqrt{1-b^2 u_R^2}\right)$$
$$-\frac{2aM}{b^2}\left(\sqrt{1-b^2 u_S^2} - \sqrt{1-b^2 u_R^2}\right) + \mathcal{O}\left(\frac{M^2}{b^2}\right),$$
$$\alpha_{retro} = \frac{2M}{b}\left(\sqrt{1-b^2 u_S^2} - \sqrt{1-b^2 u_R^2}\right)$$
$$+\frac{2aM}{b^2}\left(\sqrt{1-b^2 u_S^2} - \sqrt{1-b^2 u_R^2}\right) + \mathcal{O}\left(\frac{M^2}{b^2}\right).$$

If we wish to consider the deflection angle of light in such a case that the source point is closer to the receiver than the closest approach point, Equations (117) and (118) become:

$$\alpha_{prog} = \frac{2M}{b}\left(\sqrt{1-b^2 u_R^2} - \sqrt{1-b^2 u_S^2}\right)$$
$$-\frac{2aM}{b^2}\left(\sqrt{1-b^2 u_R^2} - \sqrt{1-b^2 u_S^2}\right) + \mathcal{O}\left(\frac{M^2}{b^2}\right),$$
$$\alpha_{retro} = \frac{2M}{b}\left(\sqrt{1-b^2 u_R^2} - \sqrt{1-b^2 u_S^2}\right)$$
$$+\frac{2aM}{b^2}\left(\sqrt{1-b^2 u_R^2} - \sqrt{1-b^2 u_S^2}\right) + \mathcal{O}\left(\frac{M^2}{b^2}\right).$$

8.7. Finite-Distance Corrections

In the previous subsections so far, we discussed an effect of the spin of the lens object to the deflection of light. In particular, we do not require the receiver and the source to be located at infinity. The finite-distance correction to the deflection angle of light is defined as $\delta\alpha$. This is the difference between the asymptotic deflection angle α_∞ and the deflection angle for the finite distance case. Namely:

$$\delta\alpha \equiv \alpha - \alpha_\infty. \tag{123}$$

Equations (117) and (118) tell us the magnitude of the finite-distance correction to the gravitomagnetic bending angle due to the spin. The result is:

$$|\delta\alpha_{GM}| \sim \mathcal{O}\left(\frac{aM}{r_S^2} + \frac{aM}{r_R^2}\right)$$
$$\sim \mathcal{O}\left(\frac{J}{r_S^2} + \frac{J}{r_R^2}\right), \tag{124}$$

where $bu_R, bu_S < 1$ is assumed, $J \equiv aM$ denotes the spin angular momentum of the lens, and the subscript GM means the gravitomagnetic part. We introduce the dimensionless spin parameter as: $s \equiv a/M$. Hence, Equation (124) is rearranged as:

$$|\delta\alpha_{GM}| \sim \mathcal{O}\left(s\left(\frac{M}{r_S}\right)^2 + s\left(\frac{M}{r_R}\right)^2\right). \tag{125}$$

This implies that $\delta\alpha_{GM}$ is of the same order as the second post-Newtonian effect (with the dimensionless spin parameter).

The second-order Schwarzschild contribution to α is $15\pi M^2/4b^2$. This contribution can be obtained also by using the present method, especially by using a relation between b and r_0 in M^2 in calculating ϕ_{RS}. Appendix A provides detailed calculations at the second order of M and a. We explain detailed calculations for the integrals of K and κ_g in the present formulation. Note that $\delta\alpha_{GM}$ in the above approximations is free from the impact parameter b. We can see this fact from Figures 11–13 below.

8.8. Possible Astronomical Applications

What are possible astronomical applications? As a first example, we consider the Sun, in which its higher multipole moments are ignored for simplicity. Its spin angular momentum denoted as $J_\odot$ is $\sim 2 \times 10^{41}$ m^2 kg s^{-1} [72,73]. This means $GJ_\odot c^{-2} \sim 5 \times 10^5$ m^2, for which the dimensionless spin parameter becomes $s_\odot \sim 10^{-1}$.

Here, our assumption is that a receiver on the Earth observes the light deflected by the Sun, while the distant source is safely in the asymptotic region. For the light ray passing near the Sun, Equation (125) allows us to make an order-of-magnitude estimation of the finite-distance correction. The result is:

$$|\delta\alpha_{GM}| \sim O\left(\frac{J}{r_R^2}\right)$$
$$\sim 10^{-12} \text{arcsec.} \times \left(\frac{J}{J_\odot}\right)\left(\frac{1\text{AU}}{r_R}\right)^2, \qquad (126)$$

where $4M_\odot/R_\odot \sim 1.75$ arcsec. $\sim 10^{-5}$ rad., where M_{odot} means the solar mass and $R_\odot$ denotes the solar radius. This correction is nearly a pico-arcsecond. Therefore, the correction is beyond the reach of present and near-future technology [74,75].

Figure 11 shows the finite-distance correction to the light deflection. Our numerical calculations are consistent with the above order-of-magnitude estimation. This figure shows also the very weak dependence of $\delta\alpha$ on b.

See Figures 12 and 13 for the deflection angle with finite-distance corrections for the prograde motion and retrograde one, respectively, where we choose $r_S \sim 1.5 \times 10^8$ km and $r_R \sim \infty$. The finite-distance correction reduces the deflection angle of light. As the impact parameter b increases, the finite-distance correction also increases.

As a second example, we discuss Sgr A* that is located at our galactic center. This object is a good candidate for measuring the strong gravitational deflection of light. The distance to the receiver is much larger than the impact parameter of light. On the other hand, some of the source stars may live in our galactic center.

For Sgr A*, Equation (125) becomes:

$$|\delta\alpha_{GM}| \sim s\left(\frac{M}{r_S}\right)^2$$
$$\sim 10^{-7} \text{arcsec.} \times \left(\frac{s}{0.1}\right)\left(\frac{M}{4\times 10^6 M_\odot}\right)^2\left(\frac{0.1\text{pc}}{r_S}\right)^2, \qquad (127)$$

where we assume that the mass of the central black hole is $M \sim 4 \times 10^6 M_\odot$. This correction is nearly at a sub-microarcsecond level. Therefore, it is beyond the capability of present technology (e.g., [31–36]).

See Figure 9 for the finite-distance correction due to the source location. The result in this figure is in agreement with the above order-of-magnitude estimation. This figure suggests the very weak dependence on the impact parameter b.

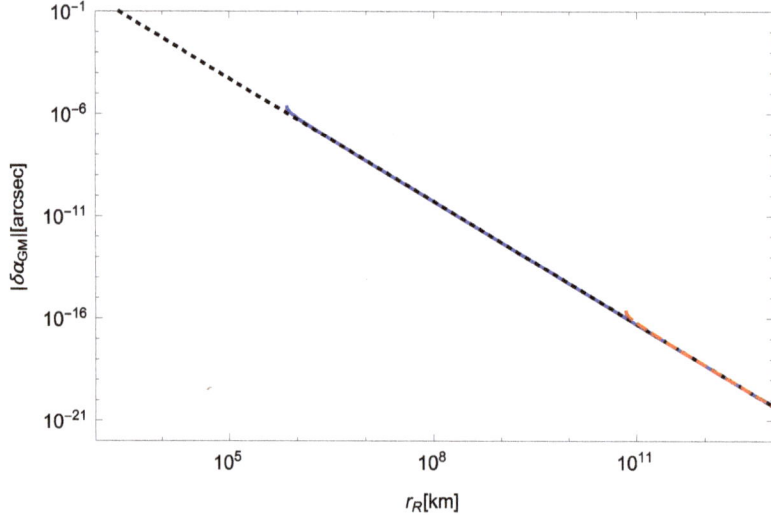

Figure 11. $\delta\alpha_{GM}$ for the Sun: The horizontal axis is the distance of the receiver distance r_R. The vertical axis means the finite-distance correction due to the gravitomagnetic deflection angle of light. The solid curve (blue in color) and dashed one (red in color) denote $b = R_\odot$ and $b = 10^5 R_\odot$, respectively. The dotted line (black in color) corresponds to the leading term in $\delta\alpha_{GM}$ given by Equation (124). These three curves are overlapped. This implies the very weak dependence of $\delta\alpha_{GM}$ on b.

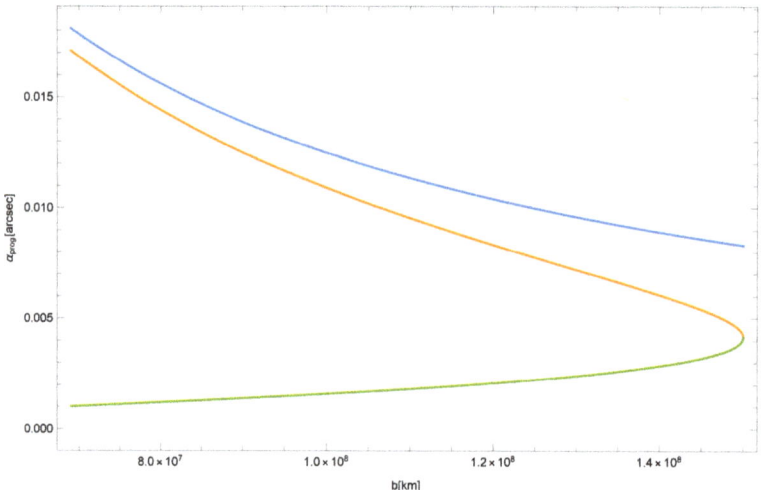

Figure 12. α in the prograde motion: The horizontal axis is the impact parameter for a photon orbit. The vertical axis means the deflection angle of light. The blue curve is the asymptotic deflection angle by a Kerr black hole. The orange curve means the deflection angle with finite-corrections by a Kerr black hole. The green curve shows the difference between the asymptotic bending angle and the deflection angle with finite-corrections by a Kerr black hole.

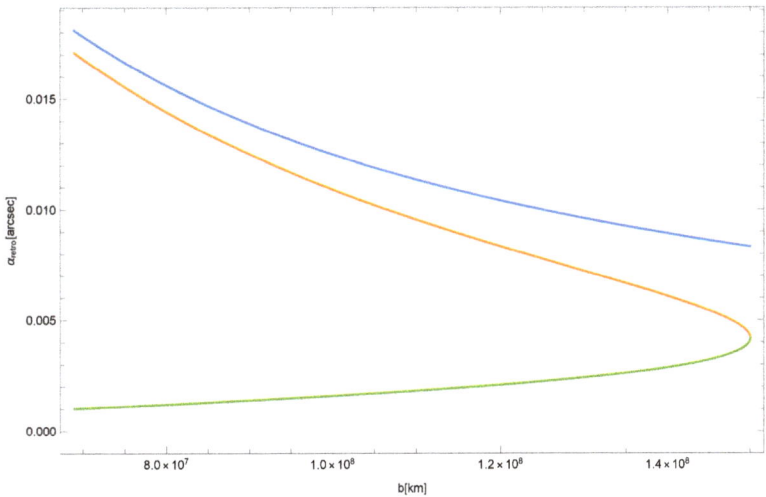

Figure 13. α for light of retrograde motion: The horizontal axis denotes the impact parameter for a photon orbit, and the vertical axis denotes the deflection angle of light. The blue curve is the asymptotic deflection angle by the Kerr black hole. The orange curve is the deflection angle with finite-correction by the Kerr black hole. The green curve shows the difference between the asymptotic bending angle and the deflection angle with finite-correction by the Kerr black hole.

9. Rotating Teo Wormhole: Another Example

9.1. Rotating Teo Wormhole and Optical Metric

In this section, we consider a rotating Teo wormhole [76] in order to examine how our method can be applied to a wormhole spacetime. The spacetime metric for this wormhole is:

$$ds^2 = -N^2 dt^2 + \frac{dr^2}{1 - \frac{b_0}{r}} + r^2 H^2 \left[d\theta^2 + \sin^2\theta (d\phi - \omega dt)^2 \right], \tag{128}$$

where we denote:

$$N = H = 1 + \frac{d(4\bar{a}\cos\theta)^2}{r}, \tag{129}$$

$$\omega = \frac{2\bar{a}}{r^3}. \tag{130}$$

Here, b_0 means the throat radius of this wormhole, $\bar{a}$ is corresponding to the spin angular momentum, and d is a positive constant.

For the rotating Teo wormhole of Equation (128), the components of the generalized optical metric are [43]:

$$\gamma_{ij} dx^i dx^j = \frac{r^7}{(r - b_0)\left(r^4 - 4\bar{a}^2 \sin^2\theta\right)\left(16 d\bar{a}^2 \cos^2\theta + r\right)^2} dr^2$$

$$+ \frac{r^6}{r^4 - 4\bar{a}^2 \sin^2\theta} d\theta^2 + \frac{r^{10} \sin^2\theta}{\left(r^4 - 4\bar{a}^2 \sin^2\theta\right)^2} d\phi^2. \tag{131}$$

Here, γ_{ij} is not the induced metric in the Arnowitt-Deser-Misner(ADM) formulation. The components of β_i are obtained as:

$$\beta_i dx^i = -\frac{2\bar{a}r^3 \sin^2\theta}{r^4 - 4\bar{a}^2 \sin^2\theta} d\phi. \tag{132}$$

In this section, we restrict ourselves within the equatorial plane, namely $\theta = \pi/2$. On the equatorial plane, the constant d in the metric always vanish because d is always associated with $\cos\theta$.

We employ the same way for the Kerr case; we first derive the orbit Equation on the equatorial plane from Equation (77) as:

$$\left(\frac{dr}{d\phi}\right)^2 = -\frac{r^5(b_0 - r)\left(4\bar{a}^2 b^2 - 4\bar{a}br^3 - b^2 r^4 + r^6\right)}{(-4\bar{a}^2 b + 2\bar{a}r^3 + br^4)^2}$$

$$= \frac{r^4}{b^2} - r^2 - \frac{b_0 r^3}{b^2} + b_0 r - \frac{4\bar{a}r^3}{b^3} + \frac{4\bar{a}b_0 r^2}{b^3} + \mathcal{O}(\bar{a}^2/b^2), \tag{133}$$

where b denotes the impact parameter of the light ray and we use the weak field and slow rotation approximations in the last line. There are no b_0 squared terms in the last line. The orbit Equation thus becomes:

$$\left(\frac{du}{d\phi}\right)^2 = \frac{1}{b^2} - u^2 - \frac{b_0 u}{b^2} + b_0 u^3 - \frac{4\bar{a}u}{b^3} - \frac{4\bar{a}b_0 u^2}{b^3} + \mathcal{O}(\bar{a}^2/b^6). \tag{134}$$

This Equation is iteratively solved as:

$$u = \frac{\sin\phi}{b} + \frac{\cos^2\phi}{2b^2} b_0 - \frac{2}{b^3}\bar{a} + \mathcal{O}\left(\frac{b_0^2}{b^3}, \frac{\bar{a}b_0}{b^4}\right). \tag{135}$$

Solving Equation (135) for ϕ_S and ϕ_R, we obtain ϕ_S and ϕ_R as:

$$\phi_S = \arcsin(bu_S) - \frac{b_0\sqrt{1 - b^2 u_S^2}}{2b} + \frac{2\bar{a}}{b^2\sqrt{1 - b^2 u_S^2}} + \mathcal{O}\left(\frac{b_0^2}{b^2}, \frac{\bar{a}b_0}{b^3}\right), \tag{136}$$

$$\phi_R = \pi - \arcsin(bu_R) + \frac{b_0\sqrt{1 - b^2 u_R^2}}{2b} - \frac{2\bar{a}}{b^2\sqrt{1 - b^2 u_R^2}} + \mathcal{O}\left(\frac{b_0^2}{b^2}, \frac{\bar{a}b_0}{b^3}\right). \tag{137}$$

9.2. Gaussian Curvature

In the weak-field approximation, the Gaussian curvature of the equatorial plane is:

$$K = -\frac{b_0}{2r^3} - \frac{56\bar{a}^2}{r^6} + \mathcal{O}\left(\frac{\bar{a}^2 b_0}{r^7}, \frac{\bar{a}^4}{r^{10}}\right), \tag{138}$$

where $\bar{a}$ and b_0 play roles as book-keeping parameters in the weak-field approximation. It is not surprising that this Gaussian curvature deviates from Equation (26) in Jusufi and Övgün [77], because their Gaussian curvature describes a different surface that is defined by using the Randers–Finsler metric. The Randers–Finsler metric is quite different from our generalized optical metric γ_{ij}.

When we perform the surface integral of the Gaussian curvature in Equation (88), we use Equation (135) for a boundary of the integration domain. The surface integral of the Gaussian curvature in Equation (88) is thus calculated as:

$$-\iint_{{}^\infty_R \square^\infty_S} K dS = \int_{\phi_S}^{\phi_R} \int_\infty^{r(\phi)} \left(-\frac{b_0}{2r^2}\right) dr d\phi + \mathcal{O}\left(\frac{b_0{}^2}{b^2}, \frac{\bar{a}b_0}{b^3}\right)$$

$$= \frac{b_0}{2} \int_{\phi_S}^{\phi_R} \int_0^{\frac{\sin\phi}{b} + \frac{\cos^2\phi}{2b^2} b_0 - \frac{2}{b^3}\bar{a}} du d\phi + \mathcal{O}\left(\frac{b_0{}^2}{b^2}, \frac{\bar{a}b_0}{b^3}\right)$$

$$= \frac{b_0}{2} \int_{\phi_S}^{\phi_R} \left[\frac{\sin\phi}{b}\right] d\phi + \mathcal{O}\left(\frac{b_0{}^2}{b^2}, \frac{\bar{a}b_0}{b^3}\right)$$

$$= \frac{b_0}{2} \left[-\frac{\cos\phi}{b}\right]_{\phi=\phi_S}^{\phi_R} + \mathcal{O}\left(\frac{b_0{}^2}{b^2}, \frac{\bar{a}b_0}{b^3}\right)$$

$$= \frac{b_0}{2b} \left(\sqrt{1-b^2 u_R{}^2} + \sqrt{1-b^2 u_S{}^2}\right) + \mathcal{O}\left(\frac{b_0{}^2}{b^2}, \frac{\bar{a}b_0}{b^3}\right), \quad (139)$$

where we use $\sin\phi_R = bu_R + \mathcal{O}(\bar{a}b^{-2}, b_0 b^{-1})$ and $\sin\phi_S = bu_S + \mathcal{O}(\bar{a}b^{-2}, b_0 b^{-1})$ by Equations (137) and (136) in the last line.

9.3. Geodesic Curvature of Photon Orbit

We study the geodesic curvature of the photon orbit on the equatorial plane in the stationary and axisymmetric spacetime by using the generalized optical metric. It generally becomes [42]:

$$\kappa_g = -\sqrt{\frac{1}{\gamma\gamma^{\theta\theta}}} \beta_{\phi,r}. \quad (140)$$

In the Teo wormhole, this expression is rearranged as:

$$\kappa_g = -\frac{2\bar{a}}{r^3} + \frac{\bar{a}b_0}{r^4} + \frac{\bar{a}b_0{}^2}{4r^5} + \frac{\bar{a}b_0{}^3}{8r^6} + \mathcal{O}\left(\frac{\bar{a}^3}{r^7}, \frac{\bar{a}^3 b_0}{r^8}\right). \quad (141)$$

We compute the path integral of the geodesic curvature of the photon orbit. The detailed calculations and result are:

$$\int_S^R \kappa_g d\ell = \int_R^S \frac{2\bar{a}}{r^3} d\ell + \mathcal{O}\left(\frac{b_0{}^2}{b^2}, \frac{\bar{a}b_0}{b^3}\right)$$

$$= \int_{\pi/2-\phi_R}^{\pi/2-\phi_S} \frac{2\bar{a}\cos\vartheta}{b^2} d\vartheta + \mathcal{O}\left(\frac{b_0{}^2}{b^2}, \frac{\bar{a}b_0}{b^3}\right)$$

$$= \frac{2\bar{a}}{b^2} \left[\sin\left(\frac{\pi}{2}-\phi_S\right) - \sin\left(\frac{\pi}{2}-\phi_R\right)\right] + \mathcal{O}\left(\frac{b_0{}^2}{b^2}, \frac{\bar{a}b_0}{b^3}\right)$$

$$= \frac{2\bar{a}}{b^2} \left(\sqrt{1-b^2 u_S{}^2} + \sqrt{1-b^2 u_R{}^2}\right) + \mathcal{O}\left(\frac{b_0{}^2}{b^2}, \frac{\bar{a}b_0}{b^3}\right), \quad (142)$$

for the retrograde orbit of the photon. In the last line, we used $\sin\phi_R = bu_R + \mathcal{O}(\bar{a}b^{-2}, b_0 b^{-1})$ and $\sin\phi_S = bu_S + \mathcal{O}(\bar{a}b^{-2}, b_0 b^{-1})$ from Equation (135). The above result becomes $4\bar{a}/b^2$, as $r_R \to \infty$ and $r_S \to \infty$. The sign of the right-hand side in Equation (142) is opposite if the photon is in prograde motion.

9.4. ϕ_{RS} Part

The rotating Teo wormhole is an asymptotically flat spacetime, as seen from Equation (128). Therefore, the integral of the geodesic curvature of the circular arc segment with an infinite radius can be expressed simply as ϕ_{RS}. By using Equations (136) and (137), ϕ_{RS} is obtained as:

$$\begin{aligned}\phi_{RS} &= \phi_R - \phi_S \\ &= \pi - \arcsin(bu_R) - \arcsin(bu_S) + \frac{b_0\sqrt{1-b^2u_R^2}}{2b} + \frac{b_0\sqrt{1-b^2u_S^2}}{2b} \\ &\quad - \frac{2\bar{a}}{b^2\sqrt{1-b^2u_R^2}} - \frac{2\bar{a}}{b^2\sqrt{1-b^2u_S^2}} + \mathcal{O}\left(\frac{b_0^2}{b^2}, \frac{\bar{a}b_0}{b^3}\right). \end{aligned} \quad (143)$$

9.5. Ψ Parts

For the rotating Teo wormhole in Equation (128), Equation (85) is computed as:

$$\sin\Psi_R = bu_R + 2\bar{a}u_R^2 - 4\bar{a}^2 bu_R^5, \quad (144)$$

and Equation (86) becomes:

$$\sin(\pi - \Psi_S) = bu_S + 2\bar{a}u_S^2 - 4\bar{a}^2 bu_S^5, \quad (145)$$

where the slow-rotation approximation is not needed.

Therefore, we obtain Ψ_R and Ψ_S as:

$$\Psi_R = \arcsin(bu_R) + \frac{2\bar{a}u_R^2}{\sqrt{1-b^2u_R^2}} + \frac{2\bar{a}^2 bu_R^5 (2b^2u_R^2 - 1)}{(b^2u_R^2 - 1)^{3/2}} + \mathcal{O}(\bar{a}^3/b^6), \quad (146)$$

$$\pi - \Psi_S = \arcsin(bu_S) + \frac{2\bar{a}u_S^2}{\sqrt{1-b^2u_S^2}} + \frac{2\bar{a}^2 bu_S^5 (2b^2u_S^2 - 1)}{(b^2u_S^2 - 1)^{3/2}} + \mathcal{O}(\bar{a}^3/b^6), \quad (147)$$

where we used the slow-rotation approximation.

9.6. Deflection Angle of Light

We combine Equations (139) and (142) to obtain the deflection angle of light in the prograde orbit as:

$$\begin{aligned}\alpha_{\text{prog}} &= \frac{b_0}{2b}\left(\sqrt{1-b^2u_R^2} + \sqrt{1-b^2u_S^2}\right) - \frac{2\bar{a}}{b^2}\left(\sqrt{1-b^2u_R^2} + \sqrt{1-b^2u_S^2}\right) \\ &\quad + \mathcal{O}\left(\frac{b_0^2}{b^2}, \frac{\bar{a}b_0}{b^3}\right). \end{aligned} \quad (148)$$

The deflection angle of the retrograde light is:

$$\begin{aligned}\alpha_{\text{retro}} &= \frac{b_0}{2b}\left(\sqrt{1-b^2u_R^2} + \sqrt{1-b^2u_S^2}\right) + \frac{2\bar{a}}{b^2}\left(\sqrt{1-b^2u_R^2} + \sqrt{1-b^2u_S^2}\right) \\ &\quad + \mathcal{O}\left(\frac{b_0^2}{b^2}, \frac{\bar{a}b_0}{b^3}\right). \end{aligned} \quad (149)$$

Next, by using Equations (143), (146), and (147), we obtain the deflection angle of the prograde light as:

$$\alpha_{\text{prog}} = \pi - \arcsin(bu_R) - \arcsin(bu_S) + \frac{b_0\sqrt{1-b^2u_R^2}}{2b} + \frac{b_0\sqrt{1-b^2u_S^2}}{2b}$$
$$- \frac{2\bar{a}}{b^2\sqrt{1-b^2u_R^2}} - \frac{2\bar{a}}{b^2\sqrt{1-b^2u_S^2}} + \arcsin(bu_R) + \frac{2\bar{a}u_R^2}{\sqrt{1-b^2u_R^2}}$$
$$- \pi + \arcsin(bu_S) + \frac{2\bar{a}u_S^2}{\sqrt{1-b^2u_S^2}} + \mathcal{O}\left(\frac{b_0^2}{b^2}, \frac{\bar{a}b_0}{b^3}\right)$$
$$= \frac{b_0}{2b}\left(\sqrt{1-b^2u_R^2} + \sqrt{1-b^2u_S^2}\right) - \frac{2\bar{a}}{b^2}\left(\sqrt{1-b^2u_R^2} + \sqrt{1-b^2u_S^2}\right)$$
$$+ \mathcal{O}\left(\frac{b_0^2}{b^2}, \frac{\bar{a}b_0}{b^3}\right). \tag{150}$$

The deflection angle of light in the retrograde orbit is:

$$\alpha_{\text{retro}} = \frac{b_0}{2b}\left(\sqrt{1-b^2u_R^2} + \sqrt{1-b^2u_S^2}\right) + \frac{2\bar{a}}{b^2}\left(\sqrt{1-b^2u_R^2} + \sqrt{1-b^2u_S^2}\right)$$
$$+ \mathcal{O}\left(\frac{b_0^2}{b^2}, \frac{\bar{a}b_0}{b^3}\right). \tag{151}$$

The deflection of light in the prograde (retrograde) orbit is weaker (stronger) when increasing the angular momentum of the Teo wormhole. The reason is as follows. The local inertial frame in which the light travels at the light speed c in general relativity moves faster (slower). Hence, the time-of-flight of light becomes shorter (longer). On light propagation, a similar explanation is done by using the dragging of the inertial frame also by Laguna and Wolsczan [78]. They discussed the Shapiro time delay. The expression of the deflection angle of light by a rotating Teo wormhole is similar to that by Kerr black hole. This implies that it is hard to distinguish a Kerr black hole from a rotating Teo wormhole by the gravitational lens observations.

In Equations (150) and (151), the source and receiver can be located at finite distances from the wormhole. In the limit as $r_R \to \infty$ and $r_S \to \infty$, Equations (148) and (149) become:

$$\alpha_{\text{prog}} \to \frac{b_0}{b} - \frac{4\bar{a}}{b^2} + \mathcal{O}\left(\frac{b_0^2}{b^2}, \frac{\bar{a}b_0}{b^3}\right),$$
$$\alpha_{\text{retro}} \to \frac{b_0}{b} + \frac{4\bar{a}}{b^2} + \mathcal{O}\left(\frac{b_0^2}{b^2}, \frac{\bar{a}b_0}{b^3}\right). \tag{152}$$

They are in complete agreement with Equations (39) and (56) in Jusufi and Övgün [77], where they restrict themselves within the asymptotic source and receiver ($r_R \to \infty$ and $r_S \to \infty$).

9.7. Finite-Distance Corrections in the Teo Wormhole Spacetime

To be precise, we define the finite-distance correction to the deflection angle of light as the difference between the asymptotic deflection angle α_∞ and the deflection angle for the finite distance case. It is denoted as $\delta\alpha$.

We consider the following situation. An observer on the Earth sees the light deflected by the solar mass. The source of light is located in a practically asymptotic region. In other words, we choose $b_0 = M_\odot$, $\bar{a} = J_\odot$, $r_R \sim 1.5 \times 10^8$ km, $r_S \sim \infty$. See Figure 14 for the finite-distance correction due to the impact parameter b. In Figure 14, the green curve means the difference between the asymptotic bending angle and the deflection angle with finite-distance corrections, the blue curve denotes the asymptotic deflection angle, and the orange curve is the deflection angle with finite-distance corrections by the rotating Teo wormhole. The deflection angle is decreased by the finite-distance correction. If the impact parameter b increases, the finite-distance correction also increases.

See also Figure 15 for numerical calculations of the finite-distance correction due to the impact parameter b. In Figure 15, the blue curve is the deflection angle with finite-distance correction by a Kerr black hole and the red curve is the deflection angle with finite-correction by a rotating Teo wormhole. The deflection of light is stronger in a Kerr black hole case for the chosen values.

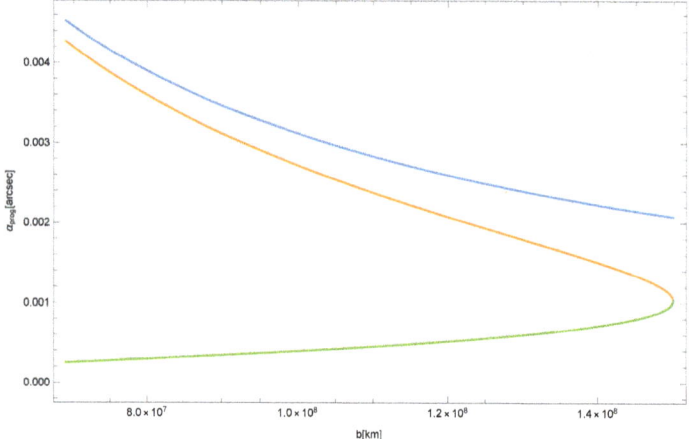

Figure 14. α in the Teo wormhole: The blue curve is the asymptotic deflection angle by the rotating Teo wormhole. The orange curve is the deflection angle with finite-distance corrections by the rotating Teo wormhole. The blue curve shows the difference between the asymptotic deflection angle and the deflection angle with finite-distance corrections by the rotating Teo wormhole.

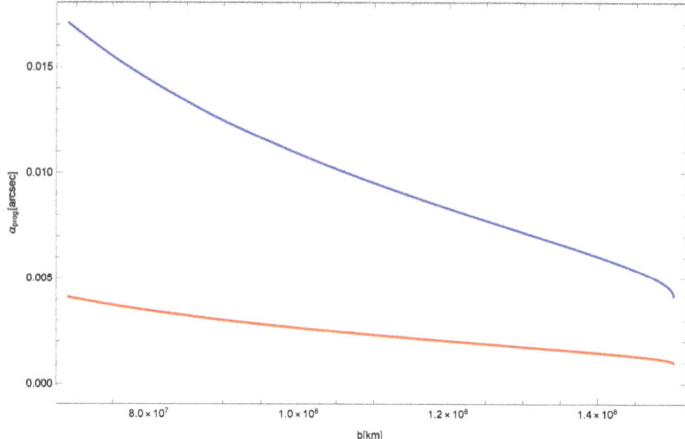

Figure 15. α for prograde motion of light: The horizontal axis is the impact parameter of photon orbit. The vertical axis means the deflection angle of light. The blue curve means the deflection angle with finite-distance corrections by the Kerr black hole. The red curve corresponds to that by the rotating Teo wormhole. For the purpose of this comparison, the mass of a Kerr black hole M and the throat radius of a rotating Teo wormhole b_0 are chosen as $M = b_0 = M_\odot$. The spin angular momentum of a Kerr black hole and that of a rotating Teo wormhole are chosen as the same as that the Sun for simplicity.

10. Summary

In this paper, we provided a brief review of a series of works on the deflection angle of light for a light source and receiver in a non-asymptotic region. [38,39,42,43]. The validity and usefulness of the new formulation come from the GB theorem in differential geometry. First, we discussed how to define the gravitational deflection angle of light in a static, spherically symmetric, and asymptotically flat spacetime, for which we assume the finite-distance source and receiver. We examined whether our definition is invariant geometrically by using the GB theorem. By using our definition, we carefully computed finite-distance corrections to the light deflection in Schwarzschild spacetime. We considered both the cases of weak deflection and the strong one. Next, we extended the definition to stationary and axisymmetric spacetimes. This extension allows us to compute finite-distance corrections for Kerr black holes and rotating Teo wormholes. We verified that these results are consistent with previous works in the infinite-distance limit. We mentioned also the finite-distance corrections to the light deflection by Sagittarius A*. It is left as future work to apply the present formulation to other interesting spacetime models and also to extend it to a more general spacetime structure.

Funding: This work was supported in part by JSPS research fellowship for young researchers (T.O.); in part by Japan Society for the Promotion of Science Grant-in-Aid for Scientific Research, No. 18J14865 (T.O.) and No. 17K05431 (H.A.); and in part by by the Ministry of Education, Culture, Sports, Science, and Technology, No. 17H06359 (H.A.).

Acknowledgments: We are grateful to Marcus Werner for the stimulating and very fruitful discussions. We thank Takao Kitamura, Asahi Ishihara, and Yusuke Suzuki for useful conversations. We would like to thank Yuuiti Sendouda, Ryuichi Takahashi, Yuya Nakamura, and Naoki Tsukamoto for useful conversations.

Conflicts of Interest: The authors declare no conflict of interest.

Appendix A. Detailed Calculations at $O(M^2/b^2)$ and $O(a^2/b^2)$ in Kerr Spacetime

First, we investigate the Gaussian curvature K of the equatorial plane in the Kerr spacetime. Here, we assume the weak-field and slow-rotation approximations. Up to the second order, K is expanded as:

$$K = \frac{R_{r\phi r\phi}}{\gamma}$$
$$= -\frac{2M}{r^3} + \frac{3M^2}{r^4} + O\left(\frac{a^2 M}{r^5}\right), \tag{A1}$$

where γ denotes det (γ_{ij}). There are no a^2 terms in K. More interestingly, only the $a^2 M$ term at the third-order level exists in K. By noting that K begins with $O(M)$, what we need for the second-order calculations is only the linear-order term in the area element on the equatorial plane. This is obtained as:

$$dS \equiv \sqrt{\gamma} dr d\phi$$
$$= \left[r + 3M + O\left(\frac{M^2}{r}\right)\right] dr d\phi, \tag{A2}$$

where terms at $O(a)$ and at $O(a^2)$ do not exist in dS. This is because all terms including the spin parameter cancel out in γ for $\theta = \pi/2$ and γ thus depends only on M, as shown by direct calculations.

By using Equations (A1) and (A2), the surface integration of the Gaussian curvature is performed as:

$$
\begin{aligned}
-\iint K dS &= \int_\infty^{r_{OE}} dr \int_{\phi_S}^{\phi_R} d\phi \left(-\frac{2M}{r^3} + \frac{3M^2}{r^4} \right)(r+3M) + O\left(\frac{M^3}{b^3}, \frac{aM^2}{b^3}, \frac{a^2M}{b^3}\right) \\
&= \int_0^{\frac{1}{b}\sin\phi + \frac{M}{b^2}(1+\cos^2\phi)} du \int_{\phi_S}^{\phi_R} d\phi \, (2M + 3uM^2) + O\left(\frac{M^3}{b^3}, \frac{aM^2}{b^3}, \frac{a^2M}{b^3}\right) \\
&= \int_{\phi_S}^{\phi_R} \left[\frac{2M}{b}\sin\phi + \frac{M^2}{2b^2}(7+\cos^2\phi)\right] d\phi + O\left(\frac{M^3}{b^3}, \frac{aM^2}{b^3}, \frac{a^2M}{b^3}\right) \\
&= \frac{2M}{b}\left[\cos\phi\right]_{\phi_R}^{\phi_S} + \frac{M^2}{2b^2}\left[\frac{30\phi + \sin(2\phi)}{4}\right]_{\phi_S}^{\phi_R} + O\left(\frac{M^3}{b^3}, \frac{aM^2}{b^3}, \frac{a^2M}{b^3}\right) \\
&= \frac{2M}{b}\left[\sqrt{1-b^2 u_S^2} + \sqrt{1-b^2 u_R^2}\right] \\
&\quad + \frac{15M^2}{4b^2}[\pi - \arcsin(bu_S) - \arcsin(bu_R)] \\
&\quad + \frac{M^2}{4b^2}\left[\frac{bu_S(15 - 7b^2 u_S^2)}{\sqrt{1-b^2 u_S^2}} + \frac{bu_R(15 - 7b^2 u_R^2)}{\sqrt{1-b^2 u_R^2}}\right] + O\left(\frac{M^3}{b^3}, \frac{aM^2}{b^3}, \frac{a^2M}{b^3}\right),
\end{aligned} \quad (A3)
$$

where we use, in the second line, an iterative solution for the orbit equation of Equation (77) in the Kerr spacetime.

Next, we study the geodesic curvature. On the equatorial plane, we find:

$$
\begin{aligned}
\kappa_g &= -\frac{1}{\sqrt{\frac{\Sigma^2}{\Delta(\Sigma-2Mr)}}\left(r^2 + a^2 + \frac{2a^2 Mr \sin^2\theta}{\Sigma}\right) \frac{\Sigma \sin^2\theta}{(\Sigma-2Mr)}} \beta_{\phi,r} \\
&= -\frac{2aM}{r^3} + O\left(\frac{aM^2}{r^3}\right).
\end{aligned} \quad (A4)
$$

Note that a^2 terms do not exist. Therefore, we obtain:

$$
\begin{aligned}
\int_{C_p} \kappa_g d\ell &= -\int_S^R d\ell \left[\frac{2aM}{r^2} + O\left(\frac{aM^2}{r^3}\right)\right] \\
&= -\frac{2aM}{b^2} \int_{\phi_S}^{\phi_R} \cos\vartheta d\vartheta + O\left(\frac{aM^2}{b^3}\right) \\
&= \frac{2aM}{b^2}\left[\sqrt{1-b^2 u_R^2} + \sqrt{1-b^2 u_S^2}\right] + O\left(\frac{aM^2}{b^3}\right),
\end{aligned} \quad (A5)
$$

where we use $\sin\phi_S = \sqrt{r_S^2 - b^2}/r_S + O(M/r_S)$ and $\sin\phi_R = -\sqrt{r_R^2 - b^2}/r_R + O(M/r_R)$.

By combining Equations (A3) and (A5), we obtain:

$$\alpha \equiv -\iint_{{}_R^\infty \square_S^\infty} K dS - \int_R^S \kappa_g d\ell$$
$$= \frac{2M}{b}\left[\sqrt{1-b^2 u_S^2} + \sqrt{1-b^2 u_R^2}\right]$$
$$+ \frac{15M^2}{4b^2}\left[\pi - \arcsin(bu_S) - \arcsin(bu_R)\right]$$
$$+ \frac{M^2}{4b^2}\left[\frac{bu_S(15-7b^2 u_S^2)}{\sqrt{1-b^2 u_S^2}} + \frac{bu_R(15-7b^2 u_R^2)}{\sqrt{1-b^2 u_R^2}}\right]$$
$$- \frac{2aM}{b^2}\left[\sqrt{1-b^2 u_R^2} + \sqrt{1-b^2 u_S^2}\right] + O\left(\frac{M^3}{b^3}, \frac{aM^2}{b^3}, \frac{a^2 M}{b^3}\right). \tag{A6}$$

Note that a^2 terms and a^3 ones do not appear in α for the finite distance situation as well as in the infinite distance limit. If we assume the infinite distance limit $u_R, u_S \to 0$, Equation (A6) becomes:

$$\alpha \to \frac{4M}{b} + \frac{15\pi M^2}{4b^2} - \frac{4aM}{b^2}. \tag{A7}$$

This agrees with the known results, especially on the numerical coefficients at the order of M^2 and aM.

References

1. Einstein, A. Die Grundlage der allgemeinen Relativitatstheorie. *Ann. Phys. (Berlin)* **1916**, *49*, 769–822. [CrossRef]
2. Dyson, F.W.; Eddington, A.S.; Davidson, C. A Determination of the Deflection of Light by the Sun's Gravitational Field, from Observations Made at the Total Eclipse of May 29, 1919. *Philos. Trans. R. Soc. A* **1920**, *220*, 291. [CrossRef]
3. Hagihara, Y. Theory of the relativistic trajectories in a gravitational field of Schwarzschild. *Jpn. J. Astron. Geophys.* **1931**, *8*, 67.
4. Chandrasekhar, S. *The Mathematical Theory of Black Holes*; Oxford University Press: New York, NY, USA, 1998.
5. Misner, C.W.; Thorne, K.S.; Wheeler, J.A. *Gravitation*; Freeman: New York, NY, USA, 1973.
6. Darwin, C. The gravity field of a particle. *Proc. R. Soc. A* **1959**, *249*, 180.
7. Bozza, V. Gravitational lensing in the strong field limit. *Phys. Rev. D* **2002**, *66*, 103001. [CrossRef]
8. Iyer, S.V.; Petters, A.O. Light's bending angle due to black holes: From the photon sphere to infinity. *Gen. Relativ. Gravit.* **2007**, *39*, 1563. [CrossRef]
9. Bozza, V.; Scarpetta, G. Strong deflection limit of black hole gravitational lensing with arbitrary source distances. *Phys. Rev. D* **2007**, *76*, 083008. [CrossRef]
10. Frittelli, S.; Kling, T.P.; Newman, E.T. Spacetime perspective of Schwarzschild lensing. *Phys. Rev. D* **2000**, *61*, 064021. [CrossRef]
11. Virbhadra, K.S.; Ellis, G.F.R. Schwarzschild black hole lensing. *Phys. Rev. D* **2000**, *62*, 084003. [CrossRef]
12. Virbhadra, K.S. Relativistic images of Schwarzschild black hole lensing. *Phys. Rev. D* **2009**, *79*, 083004. [CrossRef]
13. Virbhadra, K.S.; Narasimha, D.; Chitre, S.M. Role of the scalar field in gravitational lensing. *Astron. Astrophys.* **1998**, *337*, 1.
14. Virbhadra, K.S.; Ellis, G.F.R. Gravitational lensing by naked singularities. *Phys. Rev. D* **2002**, *65*, 103004. [CrossRef]
15. Virbhadra, K.S.; Keeton, C.R. Time delay and magnification centroid due to gravitational lensing by black holes and naked singularities. *Phys. Rev. D* **2008**, *77*, 124014. [CrossRef]
16. Zschocke, S. A generalized lens equation for light deflection in weak gravitational fields. *Class. Quantum Gravity* **2011**, *28*, 125016. [CrossRef]

17. Eiroa, E.F.; Romero, G.E.; Torres, D.F. Reissner-Nordstrom black hole lensing. *Phys. Rev. D* **2002**, *66*, 024010. [CrossRef]
18. Perlick, V. Exact gravitational lens equation in spherically symmetric and static spacetimes. *Phys. Rev. D* **2004**, *69*, 064017. [CrossRef]
19. Abe, F. Gravitational Microlensing by the Ellis Wormhole. *Astrophys. J.* **2010**, *725*, 787. [CrossRef]
20. Toki, Y.; Kitamura, T.; Asada, H.; Abe, F. Astrometric Image Centroid Displacements due to Gravitational Microlensing by the Ellis Wormhole. *Astrophys. J.* **2011**, *740*, 121. [CrossRef]
21. Nakajima, K.; Asada, H. Deflection angle of light in an Ellis wormhole geometry. *Phys. Rev. D* **2012**, *85*, 107501. [CrossRef]
22. Gibbons, G.W.; Vyska, M. The Application of Weierstrass elliptic functions to Schwarzschild Null Geodesics. *Class. Quant. Grav.* 2012, *29*, 065016. [CrossRef]
23. DeAndrea, J.P.; Alexander, K.M. Negative time delay in strongly naked singularity lensing. *Phys. Rev. D* **2014**, *89*, 123012. [CrossRef]
24. Kitamura, T.; Nakajima, K.; Asada, H. Demagnifying gravitational lenses toward hunting a clue of exotic matter and energy. *Phys. Rev. D* **2013**, *87*, 027501. [CrossRef]
25. Tsukamoto, N.; Harada, T. Signed magnification sums for general spherical lenses. *Phys. Rev. D* **2013**, *87*, 024024. [CrossRef]
26. Izumi, K.; Hagiwara, C.; Nakajima, K.; Kitamura, T.; Asada, H. Gravitational lensing shear by an exotic lens object with negative convergence or negative mass. *Phys. Rev. D* **2013**, *88*, 024049. [CrossRef]
27. Kitamura, T.; Izumi, K.; Nakajima, K.; Hagiwara, C.; Asada, H. Microlensed image centroid motions by an exotic lens object with negative convergence or negative mass. *Phys. Rev. D* **2014**, *89*, 084020. [CrossRef]
28. Nakajima, K.; Izumi, K.; Asada, H. Negative time delay of light by a gravitational concave lens. *Phys. Rev. D* **2014**, *90*, 084026. [CrossRef]
29. Tsukamoto, N.; Kitamura, T.; Nakajima, K.; Asada, H. Gravitational lensing in Tangherlini spacetime in the weak gravitational field and the strong gravitational field. *Phys. Rev. D* **2014**, *90*, 064043. [CrossRef]
30. Azreg-Ainou, M. Confined-exotic-matter wormholes with no gluing effects–Imaging supermassive wormholes and black holes. *J. Cosmol. Astropart. Phys.* **2015**, *07*, 037. [CrossRef]
31. Akiyama, K.; Alberdi, A.; Alef, W.; Asada, K.; Azulay, R.; Baczko, A.K.; Ball1, D.; Baloković, M.; Barrett, J.; Bintley, D.; et al. [Event Horizon Telescope Collaboration]. First M87 Event Horizon Telescope Results. I. The Shadow of the Supermassive Black Hole. *Astrophys. J.* 2019, *875*, L1.
32. Akiyama, K.; Alberdi, A.; Alef, W.; Asada, K.; Azulay, R.; Baczko, A.K.; Ball1, D.; Baloković, M.; Barrett, J.; Bintley, D.; et al. [Event Horizon Telescope Collaboration]. First M87 Event Horizon Telescope Results. II. Array and Instrumentation. *Astrophys. J.* 2019, *875*, L2.
33. Akiyama, K.; Alberdi, A.; Alef, W.; Asada, K.; Azulay, R.; Baczko, A.K.; Ball1, D.; Baloković, M.; Barrett, J.; Bintley, D.; et al. [Event Horizon Telescope Collaboration]. First M87 Event Horizon Telescope Results. III. Data Processing and Calibration. *Astrophys. J.* 2019, *875*, L3.
34. Akiyama, K.; Alberdi, A.; Alef, W.; Asada, K.; Azulay, R.; Baczko, A.K.; Ball1, D.; Baloković, M.; Barrett, J.; Bintley, D.; et al. [Event Horizon Telescope Collaboration]. First M87 Event Horizon Telescope Results. IV. Imaging the Central Supermassive Black Hole. *Astrophys. J.* 2019, *875*, L4.
35. Akiyama, K.; Alberdi, A.; Alef, W.; Asada, K.; Azulay, R.; Baczko, A.K.; Ball1, D.; Baloković, M.; Barrett, J.; Bintley, D.; et al. [Event Horizon Telescope Collaboration]. First M87 Event Horizon Telescope Results. V. Physical Origin of the Asymmetric Ring. *Astrophys. J.* 2019, *875*, L5.
36. Akiyama, K.; Alberdi, A.; Alef, W.; Asada, K.; Azulay, R.; Baczko, A.K.; Ball1, D.; Baloković, M.; Barrett, J.; Bintley, D.; et al. [Event Horizon Telescope Collaboration]. First M87 Event Horizon Telescope Results. VI. The Shadow and Mass of the Central Black Hole. *Astrophys. J.* 2019, *875*, L6.
37. Gibbons, G.W.; Werner, M.C. Applications of the Gauss-Bonnet theorem to gravitational lensing. *Class. Quant. Grav.* **2008**, *25*, 235009. [CrossRef]
38. Ishihara, A.; Suzuki, Y.; Ono, T.; Kitamura, T.; Asada, H. Gravitational bending angle of light for finite distance and the Gauss-Bonnet theorem. *Phys. Rev. D* **2016**, *94*, 084015. [CrossRef]
39. Ishihara, A.; Suzuki, Y.; Ono, T.; Asada, H. Finite-distance corrections to the gravitational bending angle of light in the strong deflection limit. *Phys. Rev. D* **2017**, *95*, 044017. [CrossRef]
40. Arakida, H. Light deflection and Gauss-Bonnet theorem: Definition of total deflection angle and its applications. *Gen. Rel. Grav.* **2018**, *50*, 48. [CrossRef]

41. Crisnejo, G.; Gallo, E.; Rogers, A. Finite distance corrections to the light deflection in a gravitational field with a plasma medium. *Phys. Rev. D* **2019**, *99*, 124001. [CrossRef]
42. Ono, T.; Ishihara, A.; Asada, H. Gravitomagnetic bending angle of light with finite-distance corrections in stationary axisymmetric spacetimes. *Phys. Rev. D* **2017**, *96*, 104037. [CrossRef]
43. Ono, T.; Ishihara, A.; Asada, H. Deflection angle of light for an observer and source at finite distance from a rotating wormhole. *Phys. Rev. D* **2018**, *98*, 044047. [CrossRef]
44. Ovgun, A. Light deflection by Damour-Solodukhin wormholes and Gauss-Bonnet theorem. *Phys. Rev. D* **2018**, *98*, 044033. [CrossRef]
45. Ono, T.; Ishihara, A.; Asada, H. Deflection angle of light for an observer and source at finite distance from a rotating global monopole. *Phys. Rev. D* **2019**, *99*, 124030. [CrossRef]
46. Carmo, M.P.D. *Differential Geometry of Curves and Surfaces*; Prentice-Hall: Upper Saddle River, NJ, USA, 1976; pp. 268–269.
47. Kottler, F. Uber die physikalischen Grundlagen der Einsteinschen Gravitationstheorie. *Annalen. Phys.* **1918**, *361*, 401–462. [CrossRef]
48. Bach, R. Zur Weylschen Relativitatstheorie und der Weylschen Erweiterung des Krummungstensorbegriffs. *Math. Zeit.* **1921**, *9*, 110–135. [CrossRef]
49. Riegert, R.J. Birkhoff's Theorem in Conformal Gravity. *Phys. Rev. Lett.* **1984**, *53*, 315. [CrossRef]
50. Mannheim, P.D.; Kazanas, D. Exact vacuum solution to conformal Weyl gravity and galactic rotation curves. *Astrophys. J.* **1989**, *342*, 635–638. [CrossRef]
51. Edery, A.; Paranjape, M.B. Classical tests for Weyl gravity: Deflection of light and time delay. *Phys. Rev. D* **1998**, *58*, 024011. [CrossRef]
52. Sultana, J.; Kazanas, D. Bending of light in conformal Weyl gravity. *Phys. Rev. D* **2010**, *81*, 127502. [CrossRef]
53. Cattani, C.; Scalia, M.; Laserra, E.; Bochicchio, I.; Nandi, K.K. Correct light deflection in Weyl conformal gravity. *Phys. Rev. D* **2013**, *87*, 047503. [CrossRef]
54. Sereno, M. Role of Λ in the Cosmological Lens Equation. *Phys. Rev. Lett.* **2009**, *102*, 021301. [CrossRef] [PubMed]
55. Bhadra, A.; Biswas, S.; Sarkar, K. Gravitational deflection of light in the Schwarzschild–de Sitter space-time. *Phys. Rev. D* **2010**, *82*, 063003. [CrossRef]
56. Arakida, H.; Kasai, M. Effect of the cosmological constant on the bending of light and the cosmological lens equation. *Phys. Rev. D* **2012**, *85*, 023006. [CrossRef]
57. Lim, Y.; Wang, Q. Exact gravitational lensing in conformal gravity and Schwarzschild–de Sitter spacetime. *Phys. Rev. D* **2017**, *95*, 024004. [CrossRef]
58. Lewis, T. Some Special Solutions of the Equations of Axially Symmetric Gravitational Fields. *Proc. Roy. Soc. A* **1932**, *136*, 176. [CrossRef]
59. Levy, H.; Robinson, W.J. The rotating body problem. *Proc. Camb. Philos. Soc.* **1964**, *60*, 279. [CrossRef]
60. Papapetrou, A. Champs gravitationnels stationnaires a symetrie axiale. *Ann. Inst. H. Poincare A* **1966**, *4*, 83–105.
61. Levi-Civita, T. *Absolute Differential Calculus*; Blackie and Son: Glasgow, UK, 1927.
62. Asada, H.; Kasai, M. Can We See a Rotating Gravitational Lens? *Prog. Theor. Phys.* **2000**, *104*, 95–102. [CrossRef]
63. Belton, A.C. *Geometry of Curves and Surfaces*; 2015; p. 38. Available online: www.maths.lancs.ac.uk/~belton/www/notes/geom_notes.pdf (accessed on 10 June 2019).
64. Oprea, J. *Differential Geometry and Its Applications*, 2nd ed.; Prentice Hall: Upper Saddle River, NJ, USA, 2003; p. 210.
65. Perlick, V. *Ray Optics, Fermat's Principle, and Applications to General Relativity*; Springer: Berlin, Germany, 2000.
66. Kopeikin, S.; Mashhoon, B. Gravitomagnetic effects in the propagation of electromagnetic waves in variable gravitational fields of arbitrary-moving and spinning bodies. *Phys. Rev. D* **2002**, *65*, 064025. [CrossRef]
67. Werner, M.C. Gravitational lensing in the Kerr-Randers optical geometry. *Gen. Rel. Grav.* **2012**, *44*, 3047–3057. [CrossRef]
68. Epstein, R.; Shapiro, I.I. Post-post-Newtonian deflection of light by the Sun. *Phys. Rev. D* **1980**, *22*, 2947. [CrossRef]
69. Ibanez, J. Gravitational lenses with angular momentum. *Astron. Astrophys.* **1983**, *124*, 175–180.

70. Iyer, S.V.; Hansen, E.C. Light's bending angle in the equatorial plane of a Kerr black hole. *Phys. Rev. D* **2009**, *80*, 124023. [CrossRef]
71. Kraniotis, G.V. Precise analytic treatment of Kerr and Kerr-(anti) de Sitter black holes as gravitational lenses. *Class. Quant. Grav.* **2011**, *28*, 085021.
72. Pijpers, F.P. Helioseismic determination of the solar gravitational quadrupole moment. *Mon. Not. R. Astron. Soc.* **1998**, *297*, L76. [CrossRef]
73. Bi, S.L.; Li, T.D.; Li, L.H.; Yang, W.M. Solar Models with Revised Abundance. *Astrophys. J. Lett.* **2011**, *731*, L42. [CrossRef]
74. Gaia. Available online: http://sci.esa.int/gaia/ (accessed on 10 June 2019).
75. JASMINE. Available online: http://www.jasmine-galaxy.org/index-en.html (accessed on 10 June 2019).
76. Teo, E. Rotating traversable wormholes. *Phys. Rev. D* **1998**, *58*, 024014. [CrossRef]
77. Jusufi, K.; Övgün, A. Exact traversable wormhole solution in bumblebee gravity. *Phys. Rev. D* **2018**, *97*, 024042. [CrossRef]
78. Laguna, P.; Wolszczan, A. Pulse Arrival Times from Binary Pulsars with Rotating Black Hole Companions. *Astrophys. J.* **1997**, *486*, L27. [CrossRef]

 © 2019 by the authors. Licensee MDPI, Basel, Switzerland. This article is an open access article distributed under the terms and conditions of the Creative Commons Attribution (CC BY) license (http://creativecommons.org/licenses/by/4.0/).

Article

Deflection Angle of Photons through Dark Matter by Black Holes and Wormholes Using Gauss–Bonnet Theorem

Ali Övgün [1,2]

[1] Instituto de Física, Pontificia Universidad Católica de Valparaíso, Casilla 4950, Valparaíso, Chile; ali.ovgun@pucv.cl
[2] Physics Department, Arts and Sciences Faculty, Eastern Mediterranean University, North Cyprus via Mersin 10, Famagusta 99628, Turkey; ali.ovgun@emu.edu.tr

Received: 17 March 2019; Accepted: 9 May 2019; Published: 14 May 2019

Abstract: In this research, we used the Gibbons–Werner method (Gauss–Bonnet theorem) on the optical geometry of a black hole and wormhole, extending the calculation of weak gravitational lensing within the Maxwell's fish eye-like profile and dark-matter medium. The angle is seen as a partially topological effect, and the Gibbons–Werner method can be used on any asymptotically flat Riemannian optical geometry of compact objects in a dark-matter medium.

Keywords: gravitational lensing; weak deflection; dark matter; Gauss–Bonnet theorem; black hole; wormhole

PACS: 95.30.Sf; 98.62.Sb; 97.60.Lf

1. Introduction

Black holes are an essential component of our universe, and one of the most important discoveries in astrophysics is that, when stars die, they can collapse to become extremely small objects. Black holes provide an important opportunity for probing and testing the fundamental laws of the universe. For example, gravitational waves from black holes and neutron star mergers have been recently detected [1]. Black holes may also hint at the nature of quantum gravity at small scales that change the area law of entropy. Quantum gravity is far from understood, though theoretically it has seen tremendous progress, and, in a few years, the Event Horizon Telescope may provide some more information about it [2–5].

In 1854, Maxwell presented the solution to a mathematical problem related to the passage of rays through a sphere of variable refractive index, and he noted that the potential existence of a medium of this kind would possess exceptional optical properties [6]. This is similar to the reflection of the crystalline lens in fish. This optical tool is Maxwell's fish eye (MFE), the condition in which all light rays form circular trajectories. It was a remarkable accomplishment to visualize that a lens whose refractive index increases toward a point could form perfect images [7].

Luneburg discovered that the ray propagation of MFE is equivalent to ray propagation on a homogeneous sphere with a unit radius and a unit refractive index within geometrical optics [8]. This showed that the imaging of variants that have been applied to microwave devices and the fish-eye lens in photography form an extremely wide-angled image, almost hemispherical in coverage. In 2009, Leonhardt showed that MFE is also good for waves, and it enables the production of super-resolution imaging with perfect lensing, which requires negative refractive-index materials. This began a debate and offered a rich area of research to explore [9–11]. It was shown that perfect imaging in a homogeneous three-dimensional region is also possible [12]. MFE happens when all light rays arising from any point within converge at its conjugate, which means that power released from a source can only be fully absorbed at its image point, resulting in perfect imaging. There has been a rapid increase in the importance of perfect imaging in theoretical and experimental optics [13–15].

Fermat's principle says that light rays always follow extremal optical paths, with path length being measured by refractive index n. The formula for MFE indicates the interesting possibility that rays generate a perfect image in a black hole. The refractive index depends only on distance r from the origin [13,14]. In this paper, we try to understand the effect of an MFE-like profile on the deflection angle. For simplicity, we used the uniform MFE-like profile, which is different from a nonuniform MFE profile.

Gravitational lensing is a useful tool of astronomy and astrophysics [16], in which light rays from distant stars and galaxies are deflected by a planet, a black hole, or dark matter [17,18]. The detection of dark-matter filaments [19] using weak deflection is a very relevant topic because it can help in understanding the large-scale structure of the universe [20]. To build a sky map (the refractive index of the entire visible universe), there is ongoing research on the observation of the effect of cosmological weak deflection on temperature fluctuations in the cosmic microwave background (CMB) [21]. From a theoretical point of view, new methods have been proposed to calculate deflection angle. One of them is the Gauss–Bonnet theorem (GBT), which was first proposed by Gibbons and Werner using optical geometry [22–24]. The deflection angle is seen as a partially topological effect that can be calculated by integrating the Gaussian curvature of the optical metric outward from the light ray by using the following equation: [22,23]

$$\hat{\alpha} = -\int\int_{D_\infty} K dA. \tag{1}$$

Since Gibbons and Werner's paper on weak deflection angles by GBT provided a unique perspective, this method has been applied in various cases [25–47].

Dark matter makes up to 27% of the total mass–energy of the universe [48]. We can only detect dark matter from its gravitational interactions, and we only know that dark matter is nonbaryonic, nonrelativistic (or cold), and it has weak nongravitational interactions. There are many dark-matter candidates, such as weakly interacting massive particles (WIMPs), super-WIMPs, axions, and sterile neutrinos [49]. It has been proposed that dark matter is a composite, such as the dark-atom model, which we investigate here using the deflection of light through it. Dark matter, although suppressed, generally has electromagnetic interactions [50], such that the medium of dark matter should have some optical properties that a traveling photon can sense because of the frequency-dependent refractive index. The refractive index regulates the speed at which a wave is propagated via a medium. The particles of dark matter do not get electrically charged, but they can couple to other particles that have a virtual electromagnetic charge, and can also couple to photons [51–54]. To find the amplitude of dark-matter annihilation into two photons, we must first calculate the scattering amplitude. One can obtain the index of the refractive of light, where the real part is related to the speed of propagation.

To investigate weak deflection through dark matter, we consider the propagation effects for the case that particles of dark matter (warm thermal relics or axionlike particles) have low mass whose number density is larger than ordinary matter. Put simply, dark matter interacts with photons (if only through quantum fluctuations), resulting in a refractive index. The relationship between refractive index and the forward Compton amplitude at relatively low photon energies [50] ($\mathcal{M}_{\text{fwd}} \sim -\epsilon^2 e^2$) is

$$n = 1 + \frac{\rho}{4m_{dm}^2 \omega^2} \mathcal{M}_{\text{fwd}}, \tag{2}$$

where ω is the measured photon frequency, and $\rho = 1.1 \times 10^{-6}$ GeV/cm^3 is the present-day dark-matter density [50]. Neglecting spin, the amplitude is a real and even function of ω (for photon energies below the inelastic threshold); additionally, the coefficients of the $O(\omega^{2n})$ terms are positive and spin-dependent interactions can lead to odd powers in the expansion about ω. Their presence could give information on the spin of dark matter. Hence, the refractive index becomes [50]:

$$n = 1 + \frac{\rho}{4m_{dm}^2}\left[\frac{A}{\omega^2} + B + C\omega^2 + \mathcal{O}\left(\omega^4\right)\right]. \tag{3}$$

To do so, we suppose that photons can be deflected through dark matter due to dispersive effects. We used the index of refractive $n(\omega)$ that is manipulated by the scattering amplitude of the light and dark matter in the forward [50].

Gravitational lensing in plasma has been studied in various cases [45,46,55–57]. For the first time, Bisnovatyi-Kogan and Latimer showed that, due to the dispersive properties of plasma, even in homogeneous plasma, gravitational deflection differs from a vacuum deflection angle [56,57]. Moreover, it was shown that the deflection angle is increased due to the presence of plasma [55]. Afterward, Crisnejo and Gallo calculated weak lensing in a plasma medium using the GBT [45].

The main motivation of this research is to shed light on the unexpected features of spacetimes in regards to an MFE-like profile, and to derive the deflection angle of black holes using the Gauss–Bonnet theorem in weak limits for a dark-matter medium. We suppose that the refractive index of the medium is spatially nonuniform but it is uniform at large distances. We also investigated the effect of various parameters on the refractive index of the medium, which has not been covered in previous studies.

2. Effect of Medium on Deflection Angle of Schwarzschild Black Hole Using Gauss–Bonnet Theorem

In this section, we first describe the black-hole solution in a static and spherically symmetric spacetime. Then, we apply the MFE-like profile within the GBT to calculate the weak deflection angle.

The Schwarzschild black-hole spacetime reads

$$ds^2 = -f(r)dt^2 + g(r)dr^2 + r^2(d\theta^2 + \sin^2\theta d\varphi^2), \tag{4}$$

with metric functions

$$f(r) = g(r)^{-1} = 1 - \frac{2M}{r}. \tag{5}$$

Analysis of the geodesics equation, the ray equation, is calculated by

$$\varphi = \int \frac{b\sqrt{g(r)}\,dr}{r^2\sqrt{\frac{1}{f(r)} - \frac{b^2}{r^2}}}, \tag{6}$$

where b is the impact parameter of the unperturbed photon.

Our universe is homogeneous and isotropic on large scales. Now, we consider isotropic coordinates that are nonsingular at the horizon and time direction is a Killing vector. Moreover, time slices become Euclidean with a conformal factor, and one can calculate refractive index n of light rays around the black hole. Another important feature of isotropic coordinates is that they satisfy Landau's condition of coordinate clock synchronization:

$$\frac{\partial}{\partial x_j}\left(-\frac{g_{0i}}{g_{00}}\right) = \frac{\partial}{\partial x_i}\left(-\frac{g_{0j}}{g_{00}}\right) (i,j = 1,2,3). \tag{7}$$

Using transformation

$$r = \rho\left(1 + \frac{M}{2\rho}\right)^2, \tag{8}$$

the Schwarzschild black hole is rewritten in isotropic coordinates (where ρ is an isotropic radial coordinate) [45]

$$ds^2 = -F(\rho)dt^2 + G(\rho)(d\rho^2 + \rho^2 d\Omega^2), \tag{9}$$

with

$$F(\rho) = \left(\frac{\rho - \frac{M}{2}}{\rho + \frac{M}{2}}\right)^2, \text{ and } G(\rho) = \left(\frac{\rho + \frac{M}{2}}{\rho}\right)^4. \tag{10}$$

The metric becomes nonsingular at horizon $r = 2M$. It can also be written in Fermat form of the metric:

$$ds^2 = F(\rho)[-dt^2 + n(\rho)^2(d\rho^2 + \rho^2 d\Omega^2)], \tag{11}$$

with index of refractive $n(\rho) = \frac{c}{v(\rho)}$. For the Schwarzschild black-hole medium, the refractive index reads

$$n = \frac{\left(1 + \frac{M}{2\rho}\right)^3}{\left(1 - \frac{M}{2\rho}\right)}, \tag{12}$$

and it can be approximated for large $\rho \gg M$

$$n \approx 1 + \frac{2M}{\rho}. \tag{13}$$

Now, the ray equation becomes

$$\varphi = \int \frac{b \, d\rho}{\rho^2 \sqrt{n^2 - \frac{b^2}{\rho^2}}}. \tag{14}$$

To discuss the deflection angle and extract information of the MFE-like profile, the GBT was used instead of the null geodesics method. The GBT is calculated using the negative Gauss curvature of the optical metric.

2.1. Case 1

Let us start from the constant case for medium n_m as refractive index:

$$n_m = n_0, \tag{15}$$

where n_0 is a constant refractive index of the medium; here, we consider the GBT to obtain the deflection angle in a medium in weak field limits.

Let us write the optical Schwarzschild spacetime in an equatorial plane [45]:

$$d\sigma^2 = \frac{n_m^2}{f(\rho)}[g(\rho)d\rho^2 + \rho^2 d\varphi^2]. \tag{16}$$

Then, we calculate the Gaussian optical curvature:

$$K = -2\frac{M}{n_0^2 \rho^3} + 3\frac{M^2}{n_0^2 \rho^4} + O(M^3), \tag{17}$$

which is negative everywhere that gives a universal property of black-hole metrics [23].
It reduces to this form at a linear order of M:

$$K \approx -2\frac{M}{n_0^2 \rho^3}. \tag{18}$$

This result is used to evaluate the deflection angle using a nonsingular domain outside the light ray (D_ρ, with boundary $\partial D_\rho = \gamma \cup C_\rho$) [22]:

$$\iint_{D_\rho} K\, dS + \oint_{\partial D_\rho} \kappa\, dt + \sum_i \theta_i = 2\pi \chi(D_\rho), \quad (19)$$

where κ stands for the geodesic curvature, and K is the Gaussian optical curvature, with exterior angles $\theta_i = (\theta_O, \theta_S)$ and Euler characteristic number $\chi(D_\rho) = 1$. At weak limits, ($\rho \to \infty$), $\theta_O + \theta_S \to \pi$. Then, the GBT reduces to

$$\iint_{D_\rho} K\, dS + \oint_{C_\rho} \kappa\, dt \stackrel{\rho \to \infty}{=} \iint_{D_\infty} K\, dS + \int_0^{\pi + \hat{\alpha}} d\varphi = \pi. \quad (20)$$

For geodesics γ, geodesic curvature vanishes $\kappa(\gamma) = 0$, and we have

$$\kappa(C_\rho) = |\nabla_{\dot{C}_\rho} \dot{C}_\rho|, \quad (21)$$

with $C_\rho = \rho$ = constant. The GBT becomes

$$\lim_{\rho \to \infty} \int_0^{\pi + \hat{\alpha}} \left[\kappa \frac{d\sigma}{d\varphi}\right]\bigg|_{C_\rho} d\varphi = \pi - \lim_{\rho \to \infty} \iint_{D_\rho} K\, dS, \quad (22)$$

and one can calculate

$$\frac{d\sigma}{d\varphi}\bigg|_{C_\rho} = n_m \left(\frac{\rho^3}{\rho - 2M}\right)^{1/2}, \quad (23)$$

where, for very large radial distance,

$$\kappa(C_\rho) dt = d\varphi. \quad (24)$$

Therefore, as expected, for this number density profile and physical metric (which imply that the optical metric is asymptotically Euclidean), we corroborate that

$$\lim_{\rho \to \infty} \kappa \frac{d\sigma}{d\varphi}\bigg|_{C_\rho} = 1. \quad (25)$$

At a linear order in M, it follows to use Equation (22) in limit $\rho \to \infty$, and taking geodesic curve γ, approximated by its flat Euclidean version parametrized as $\rho = b/\sin\varphi$, with b representing the impact parameter in the physical spacetime that

$$\hat{\alpha} = -\lim_{\rho \to \infty} \int_0^\pi \int_{\frac{b}{\sin\varphi}}^\rho K\, dS. \quad (26)$$

After nontrivial calculation, we calculate that the deflection angle of the Schwarzschild black hole in medium for the leading order terms is

$$\hat{\alpha} = 4\frac{M}{n_0 b}, \quad (27)$$

which agrees with the well-known results in the limit at which its presence is negligible ($n_0 = 1$); this expression reduces to known vacuum formula $\hat{\alpha} = 4\frac{M}{b}$., so that GBT exhibits a partially topological effect. This method can be used in any asymptotically flat Riemannian optical metrics.

2.2. Case 2

Now, we apply the different model of the MFE-like medium [10]

$$n = \frac{z_0}{1 + z^2}, \quad (28)$$

where z_0 and z are a constant.

The Gaussian curvature of the optical metric approximating in leading orders is negative everywhere and found as:

$$K = -2\frac{(z^2+1)^2 M}{z_0^2 \rho^3} + O(M^3), \tag{29}$$

Then, using the same method, we calculate the deflection angle as follows:

$$\hat{\alpha} \simeq 4\frac{Mz^2}{z_0 b} + 4\frac{M}{z_0 b}. \tag{30}$$

At $z = 0$ and $z_0 = 1$, it reduces to the exact Schwarzschild case.

2.3. Case 3

The refractive index for the dark-matter medium [50]

$$n(\omega) = 1 + \beta A_0 + A_2 \omega^2 \tag{31}$$

where $\beta = \frac{\rho_0}{4m^2\omega^2}$ and ρ_0 are the mass density of the scattered dark-matter particles, and $A_0 = -2\varepsilon^2 e^2$ and $A_{2j} \geq 0$.

The terms in $\mathcal{O}(\omega^2)$ and higher are related to the polarizability of the dark-matter candidate. Note that the order of ω^{-2} is due to the charged dark-matter candidate and ω^2 for a neutral dark-matter candidate. Moreover, there may be a linear term in ω when parity and charge-parity asymmetries are present.

The Gaussian curvature is obtained as:

$$K \approx -2\frac{M}{(A_2\omega^2 + \beta A_0 + 1)^2 \rho^3} + O(M^3) \tag{32}$$

The deflection angle is found as follows:

$$\hat{\alpha} = 4\frac{M}{(A_2\omega^2 + 1)b} - 4\frac{MA_0}{(A_2\omega^2+1)^2 b}\beta + O\left(\beta^2\right) \tag{33}$$

The effect of dark matter can be seen by comparison with the above deflection angle by the Schwarzschild black hole. Hence, dark matter gives a small deflection angle compared to the standard case.

3. Effect of Medium on Deflection Angle of Schwarzschild-Like Wormhole Using Gauss–Bonnet Theorem

In this section, we consider the static Schwarzschild-like wormhole solution [58] with metric

$$ds^2 = -(1 - 2M/r + \lambda^2)dt^2 + \frac{dr^2}{1 - 2M/r} + r^2 d\Omega_{(2)}^2, \tag{34}$$

which reduces to the black-hole metric in Equation (4) at $\lambda = 0$. Using the transformation of $t \to t/\sqrt{1+\lambda^2}$ and $M \to M(1+\lambda^2)$, the metric functions of the Schwarzschild-like wormhole spacetime become:

$$f(r) = 1 - \frac{2M}{r}, \quad g(r)^{-1} = 1 - \frac{2M(1+\lambda^2)}{r}. \tag{35}$$

3.1. Case 1

We first use the constant profile as refractive $n_m = n_0$ to calculate the deflection angle in the medium in weak field limits. Using the same procedure, we obtain the optical metric, and calculate the Gaussian optical curvature for the Schwarzschild-like wormhole at a linear order of M as follows:

$$K \approx -\frac{(\lambda^2 + 2) M}{\rho^3 n_0^2} + O(M^3), \tag{36}$$

and after similar calculations, the corresponding deflection angle in the leading order terms is

$$\hat{\alpha} = 2 \frac{M\lambda^2}{n_0 b} + 4 \frac{M}{n_0 b}, \tag{37}$$

which agrees with the well-known results in the limit in which the medium is negligible ($n_0 = 1$) [42].

3.2. Case 2

To see the effect of the MFE-like medium, we use [10]

$$n = \frac{z_0}{1 + z^2}, \tag{38}$$

where z_0 and z are a constant. The Gaussian curvature of the optical metric approximating in leading orders is negative everywhere and found as:

$$K \approx -\frac{(z^2 + 1)^2 (\lambda^2 + 2) M}{\rho^3 z_0^2} \tag{39}$$

Using the GBT, the deflection angle is calculated as follows:

$$\hat{\alpha} \simeq 2 \frac{M\lambda^2 z^2}{z_0 b} + 2 \frac{M\lambda^2}{z_0 b} + 4 \frac{Mz^2}{z_0 b} + 4 \frac{M}{z_0 b}. \tag{40}$$

At $z = 0$ and $z_0 = 1$, it reduces to the previous result [42].

3.3. Case 3

Finally, we use the refractive index for dark matter given in Equation (31) to calculate the deflection angle of a wormhole in a medium. The Gaussian curvature is obtained as:

$$K \approx -\frac{(\lambda^2 + 2) M}{\rho^3 (A_2 \omega^2 + A_0 \beta + 1)^2} \tag{41}$$

The deflection angle is found as follows:

$$\hat{\alpha} = 2 \frac{M\lambda^2}{(A_2 \omega^2 + A_0 \beta + 1) b} + 4 \frac{M}{(A_2 \omega^2 + A_0 \beta + 1) b} \tag{42}$$

We find that the deflected photon through dark matter around the Schwarzschild-like wormhole has a large deflection angle compared to the standard case.

4. Conclusions

We calculated the deflection angle of black holes and wormholes in a dark-matter medium using the GBT. This was achieved by constructing optical metrics. In summary, we investigated that GBT is a partially topological effect. We demonstrated this by using three different cases. In the first case, we used the constant profile as a refractive index. Then, by constructing the optical geometry and using the GBT, we obtained the deflection angle in the weak field limit. The deflection angle of the Schwarzschild black hole was correctly calculated in a medium that has a constant n_0 refractive index. In the second case, we used the MFE-like model (but uniform in large distances), repeated the calculation, and showed that it produces a similar effect.

In Section 2, we repeated our method on the Schwarzschild-like wormhole to see the effect of the dark-matter medium when light is propagated through it. Note that we supposed that the refractive index is spatially nonuniform as long as it is uniform at large distances. In the first case, we again used the constant refractive index, we considered the MFE-like profile, and, finally, the medium for the dark matter was taken to find the deflection angle in the weak field limit. We concluded that the deflection angle by a black hole decreases in a medium of dark matter, as seen in Equation (33). On the other hand, deflection angle by a wormhole increases, as seen in Equation (42).

These results suggest that weak deflection within a dark-matter medium or MFE-like model (perfect imaging), can be calculated using the Gibbons and Werner method, which gives us hints to understanding the nature of dark matter.

Funding: This work was supported by Comisión Nacional de Ciencias y Tecnología of Chile through FONDECYT Grant No. 3170035 (A.Ö.)

Acknowledgments: A.Ö. is grateful to the Institute for Advanced Study, Princeton for their hospitality.

Conflicts of Interest: The author declares no conflict of interest. The funders had no role in the design of the study; in the collection, analyses, or interpretation of data; in the writing of the manuscript; or in the decision to publish the results.

References

1. Abbott, B.P.; Abbott, R.; Abbott, T.D.; Abernathy, M.R.; Acernese, F.; Ackley, K.; Adams, C.; Adams, T.; Addesso, P.; Adhikari, R.X.; et al. Observation of Gravitational Waves from a Binary Black Hole Merger. *Phys. Rev. Lett.* **2016**, *116*, 061102. [CrossRef]
2. Nutku, Y.; Halil, M. Colliding Impulsive Gravitational Waves. *Phys. Rev. Lett.* **1977**, *39*, 1379–1382. [CrossRef]
3. Alberdi, A.; Gómez Fernández, J.L.; Event Horizon Telescope Collaboration. First M87 Event Horizon Telescope Results. I. The Shadow of the Supermassive Black Hole. Available online: http://adsabs.harvard.edu/abs/2019ApJ...875L...1E (accessed on 13 May 2019).
4. Giddings, S.B.; Psaltis, D. Event Horizon Telescope Observations as Probes for Quantum Structure of Astrophysical Black Holes. *Phys. Rev. D* **2018**, *97*, 084035. [CrossRef]
5. Barrau, A. Astrophysical and cosmological signatures of Loop Quantum Gravity. *Scholarpedia* **2017**, *12*, 33321. [CrossRef]
6. Niven, W.D. *The Scientific Papers of James Clerk Maxwell*; Maxwell's Original Piece in the Cambridge and Dublin Mathematical Journal for February 1854; Cambridge University Press: Cambridge, UK, 1890; Volume 1, pp. 76–79.
7. Pendry, J.B. Negative Refraction Makes a Perfect Lens. *Phys. Rev. Lett.* **2000**, *85*, 3966–3969. [CrossRef]
8. Luneburg, R.K. *Mathematical Theory of Optics*; Brown University: Providence, RI, USA, 1944; pp. 189–213.
9. Leonhardt, U. Perfect imaging without negative refraction. *New J. Phys.* **2009**, *11*, 093040. [CrossRef]
10. Leonhardt, U.; Sahebdivan, S. Theory of Maxwell's fish eye with mutually interacting sources and drains. *Phys. Rev. A* **2015**, *92*, 053848. [CrossRef]
11. Leonhardt, U. Optical conformal mapping. *Science* **2006**, *312*, 1777–1780. [CrossRef]
12. Minano, J.C. Perfect imaging in a homogeneous three-dimensional region. *Opt. Express* **2006**, *14*, 9627–9635. [CrossRef]
13. Tyc, T.; Danner, A. Resolution of Maxwell's fisheye with an optimal active drain. *New J. Phys.* **2014**, *16*, 063001. [CrossRef]
14. Liu, Y.; Chen, H. Infinite Maxwell fisheye inside a finite circle. *J. Opt.* **2015**, *17*, 125102. [CrossRef]
15. Guenneau, S.; Diatta, A.; McPhedran, R.C. Focusing: coming to the point in metamaterials. *J. Mod. Opt.* **2010**, *57*, 511–527. [CrossRef]
16. Perlick, V. Gravitational Lensing from a Spacetime Perspective. *Living Rev. Relat.* **2004**, *7*, 9. [CrossRef]
17. Bozza, V. Gravitational Lensing by Black Holes. *Gen. Relat. Grav.* **2010**, *42*, 2269–2300. [CrossRef]

18. Stefanov, I.Z.; Yazadjiev, S.S.; Gyulchev, G.G. Connection between Black-Hole Quasinormal Modes and Lensing in the Strong Deflection Limit. *Phys. Rev. Lett.* **2010**, *104*, 251103. [CrossRef]
19. Epps, S.D.; Hudson, M.J. The Weak Lensing Masses of Filaments between Luminous Red Galaxies. *Mon. Not. R. Astron. Soc.* **2017**, *468*, 2605–2613. [CrossRef]
20. Bartelmann, M.; Schneider, P. Weak gravitational lensing. *Phys. Rept.* **2001**, *340*, 291–472. [CrossRef]
21. Ade, P.A.; Aghanim, N.; Arnaud, M.; Ashdown, M.; Aumont, J.; Baccigalupi, C.; Banday, A.J.; Barreiro, R.B.; Bartlett, J.G.; Bartolo, N.; et al. Planck 2015 results. XIII. Cosmological parameters. *Astron. Astrophys.* **2016**, *594*, A13.
22. Gibbons, G.W.; Werner, M.C. Applications of the Gauss-Bonnet theorem to gravitational lensing. *Class. Quant. Grav.* **2008**, *25*, 235009. [CrossRef]
23. Gibbons, G.W.; Warnick, C.M. Universal properties of the near-horizon optical geometry. *Phys. Rev. D* **2009**, *79*, 064031. [CrossRef]
24. Werner, M.C. Gravitational lensing in the Kerr-Randers optical geometry. *Gen. Relat. Grav.* **2012**, *44*, 3047. [CrossRef]
25. Ishihara, A.; Suzuki, Y.; Ono, T.; Kitamura, T.; Asada, H. Gravitational bending angle of light for finite distance and the Gauss-Bonnet theorem. *Phys. Rev. D* **2016**, *94*, 084015. [CrossRef]
26. Sakalli, I.; Ovgun, A. Hawking Radiation and Deflection of Light from Rindler Modified Schwarzschild Black Hole. *Europhys. Lett.* **2017**, *118*, 60006. [CrossRef]
27. Jusufi, K.; Werner, M.C.; Banerjee, A.; Övgün, A. Light Deflection by a Rotating Global Monopole Spacetime. *Phys. Rev. D* **2017**, *95*, 104012. [CrossRef]
28. Ono, T.; Ishihara, A.; Asada, H. Gravitomagnetic bending angle of light with finite-distance corrections in stationary axisymmetric spacetimes. *Phys. Rev. D* **2017**, *96*, 104037. [CrossRef]
29. Jusufi, K.; Sakalli, I.; Övgün, A. Effect of Lorentz Symmetry Breaking on the Deflection of Light in a Cosmic String Spacetime. *Phys. Rev. D* **2017**, *96*, 024040.
30. Ishihara, A.; Suzuki, Y.; Ono, T.; Asada, H. Finitedistance corrections to the gravitational bending angle of light in the strong deflection limit. *Phys. Rev. D* **2017**, *95*, 044017. [CrossRef]
31. Jusufi, K.; Övgün, A.; Banerjee, A. Light deflection by charged wormholes in Einstein-Maxwell-dilaton theory. *Phys. Rev. D* **2017**, *96*, 084036. [CrossRef]
32. Arakida, H. Light deflection and Gauss-Bonnet theorem: Definition of total deflection angle and its applications. *Gen. Relat. Grav.* **2018**, *50*, 48. [CrossRef]
33. Jusufi, K.; Övgün, A. Gravitational Lensing by Rotating Wormholes. *Phys. Rev. D* **2018**, *97*, 024042. [CrossRef]
34. Övgün, A.; Sakalli, I.; Saavedra, J. Shadow cast and Deflection angle of Kerr-Newman-Kasuya spacetime. *J. Cosmol. Astropart. Phys.* **2018**, *1810*, 41. [CrossRef]
35. Övgün, A.; Sakalli, I.; Saavedra, J. Weak gravitational lensing by Kerr-MOG Black Hole and Gauss-Bonnet theorem. *arXiv* **2018**, arXiv:1806.06453.
36. Övgün, A.; Gyulchev, G.; Jusufi, K. Weak Gravitational lensing by phantom black holes and phantom wormholes using the Gauss-Bonnet theorem. *Ann. Phys.* **2019**, *406*, 152–172. [CrossRef]
37. Jusufi, K.; Övgün, A.; Saavedra, J.; Gonzalez, P.A.; Vasquez, Y. Deflection of light by rotating regular black holes using the Gauss-Bonnet theorem. *Phys. Rev. D* **2018**, *87*, 124024. [CrossRef]
38. Övgün, A.; Jusufi, K.; Sakalli, I. Exact traversable wormhole solution in bumblebee gravity. *Phys. Rev. D* **2019**, *99*, 024042. [CrossRef]
39. Övgün, A.; Jusufi, K.; Sakalli, I. Gravitational lensing under the effect of Weyl and bumblebee gravities: Applications of Gauss–Bonnet theorem. *Ann. Phys.* **2018**, *399*, 193–203. [CrossRef]
40. Jusufi, K.; Övgün, A. Effect of the cosmological constant on the deflection angle by a rotating cosmic string. *Phys. Rev. D* **2018**, *97*, 064030. [CrossRef]
41. Javed, W.; Babar, R.; Övgün, A. The effect of the Brane-Dicke coupling parameter on weak gravitational lensing by wormholes and naked singularities. *Phys. Rev. D* **2019**, *99*, 084012. [CrossRef]
42. Övgün, A. Light deflection by Damour-Solodukhin wormholes and Gauss-Bonnet theorem. *Phys. Rev. D* **2018**, *98*, 044033. [CrossRef]
43. Ono, T.; Ishihara, A.; Asada, H. Deflection angle of light for an observer and source at finite distance from a rotating wormhole. *Phys. Rev. D* **2018**, *98*, 044047. [CrossRef]
44. Ono, T.; Ishihara, A.; Asada, H. Deflection angle of light for an observer and source at finite distance from a rotating global monopole. *arXiv* **2018**, arXiv:1811.01739.

45. Crisnejo, G.; Gallo, E. Weak lensing in a plasma medium and gravitational deflection of massive particles using the Gauss-Bonnet theorem. A unified treatment. *Phys. Rev. D* **2018**, *97*, 124016. [CrossRef]
46. Crisnejo, G.; Gallo, E.; Rogers, A. Finite distance corrections to the light deflection in a gravitational field with a plasma medium. *arXiv* **2018**, arXiv:1807.00724.
47. Asada, H. Gravitational lensing by exotic objects. *Mod. Phys. Lett. A* **2017**, *32*, 1730031. [CrossRef]
48. Hinshaw, G.; Larson, D.; Komatsu, E.; Spergel, D.N.; Bennett, C.; Dunkley, J.; Nolta, M.R.; Halpern, M.; Hill, R.S.; Odegard, N.; et al. Nine-Year Wilkinson Microwave Anisotropy Probe (WMAP) Observations: Cosmological Parameter Results. *Astrophys. J. Suppl.* **2013**, *208*, 19. [CrossRef]
49. Feng, J.L. Dark Matter Candidates from Particle Physics and Methods of Detection. *Ann. Rev. Astron. Astrophys.* **2010**, *48*, 495. [CrossRef]
50. Latimer, D.C. Dispersive Light Propagation at Cosmological Distances: Matter Effects. *Phys. Rev. D* **2013**, *88*, 063517. [CrossRef]
51. Latimer, D.C. Anapole dark matter annihilation into photons. *Phys. Rev. D* **2017**, *95*, 095023. [CrossRef]
52. Latimer, D.C. Two-photon interactions with Majorana fermions. *Phys. Rev. D* **2016**, *94*, 093010. [CrossRef]
53. Kvam, A.K.; Latimer, D.C. Optical dispersion of composite particles consisting of millicharged constituents. *J. Phys. G* **2016**, *43*, 085002. [CrossRef]
54. Whitcomb, K.M.; Latimer, D.C. Scattering from a quantum anapole at low energies. *Am. J. Phys.* **2017**, *85*, 932. [CrossRef]
55. Tsupko, O.Y.; Bisnovatyi-Kogan, G.S. Gravitational lensing in plasma: Relativistic images at homogeneous plasma. *Phys. Rev. D* **2013**, *87*, 124009. [CrossRef]
56. Bisnovatyi-Kogan, G.S.; Tsupko, O.Y. Gravitational lensing in a non-uniform plasma. *Mon. Not. R. Astron. Soc.* **2010**, *404*, 1790–1800. [CrossRef]
57. Bisnovatyi-Kogan, G.S.; Tsupko, O.Y. Gravitational Lensing in Plasmic Medium. *Plasma Phys. Rep.* **2015**, *41*, 562–581. [CrossRef]
58. Damour, T.; Solodukhin, S.N. Wormholes as black hole foils. *Phys. Rev. D* **2007**, *76*, 024016. [CrossRef]

© 2019 by the authors. Licensee MDPI, Basel, Switzerland. This article is an open access article distributed under the terms and conditions of the Creative Commons Attribution (CC BY) license (http://creativecommons.org/licenses/by/4.0/).

Article

Shadow Images of a Rotating Dyonic Black Hole with a Global Monopole Surrounded by Perfect Fluid

Sumarna Haroon [1], Kimet Jusufi [2,3,*] and Mubasher Jamil [1,4]

1. Department of Mathematics, School of Natural Sciences (SNS), National University of Sciences and Technology (NUST), Islamabad H-12, Pakistan; sumarna.haroon@sns.nust.edu.pk (S.H.); mjamil@zjut.edu.cn (M.J.)
2. Physics Department, State University of Tetovo, Ilinden Street nn, Tetovo 1200, North Macedonia
3. Institute of Physics, Faculty of Natural Sciences and Mathematics, Ss. Cyril and Methodius University, Arhimedova 3, Skopje 1000, North Macedonia
4. Institute for Theoretical Physics and Cosmology, Zhejiang University of Technology, Hangzhou 310023, China
* Correspondence: kimet.jusufi@unite.edu.mk

Received: 29 November 2019; Accepted: 21 January 2020; Published: 24 January 2020

Abstract: In this paper, we revisit the rotating global monopole metric and extend the metric to a rotating dyonic global monopole in the presence of a perfect fluid. We then show that the surface topology at the event horizon, related to the metric computed, is a 2-sphere using the Gauss-Bonnet theorem. By choosing $\omega = -1/3, 0, 1/3$ we investigate the effect of dark matter, dust and radiation on the silhouette of a black hole. The presence of the global monopole parameter γ and the perfect fluid parameters v also deform the shape of a black hole's shadow, which has been depicted through graphical illustrations. Finally, we analyse the energy emission rate of a rotating dyonic global monopole surrounded by perfect fluid with respect to parameters.

Keywords: rotating black hole; global monopole; perfect fluid; scalar field; shadows

1. Introduction

Black holes are fascinating objects predicted to exist by Einstein's theory of general relativity. Recent astrophysical observation shows that such objects may exist at the center of almost every galaxy [1,2]. By studying the light-like geodesics around black holes it is shown that photons can be absorbed by the black hole or can escape from black holes [3]. That is to say a boundary is defined between these two categories of light-like geodesics, giving rise to a dark region known as the shadow. Very recently, a project known as the Event Horizon Telescope (EHT) Collaboration announced the first image concerning the detection of an event horizon of a supermassive black hole at the center of a giant elliptical galaxy, M87 [4,5]. That being said, the black hole shadow has recently become a hot topic among researchers for the simple fact it is best to evaluate the soon-expected observational data. Historically, Synge was the first to propose the apparent shape of a spherically symmetric black hole [6]. After that Luminet [7] discussed the appearance of a Schwarzschild black hole, the shadow of a Kerr black hole was studied by Bardeen [8], the shadow of Kerr-Newman black holes [9], naked singularities with deformation parameters [10], Kerr-Nut spacetimes [11], while shadows of black holes in Chern-Simons modified gravity, Randall-Sundrum braneworlds, and Kaluza-Klein rotating black holes have been studied in References [12–14], and many other interesting studies concerning the effect of dark matter and cosmological constant on the shadow images [15–22], Kerr-like wormholes as well as traversable wormholes and many other interesting studies [23–31]. Some authors have also tried to test theories of gravity by using the observations obtained from the shadow of Sgr A* [32–35]. Note that the new general approach for shadow calculation for axially symmetric black holes was developed in Reference [36], while the intrinsic curvature and topology of shadows in Kerr spacetime was developed in References [37,38].

Global monopoles are topological defects which may have been produced during the phase transitions in the early universe. In fact, global monopoles are just one type of topological defects. Other types of topological objects are expected to exist including domain walls and cosmic strings (e.g., [39]). More precisely, a global monopole is a heavy object characterized by spherically symmetry and divergent mass. Such objects, which may have been formed during the phase transition of a system composed of a self-coupling triplet of scalar fields ϕ^a which undergoes a spontaneous breaking of global $O(3)$, gauge symmetry down to $U(1)$. The gravitational field of a static global monopole was found by Barriola and Vilenkin for the first time and are expected to be stable against spherical as well as polar perturbations [40]. According to their model, global monopoles are configurations whose energy density decreases with the distance r^{-2} and whose spacetimes exhibit a solid angle deficit given by $\Delta = 8\pi^2\gamma^2$, where γ is the scale of gauge-symmetry breaking. The rotating metric of a global monopole was investigated by Filho and Bezerra [41] Gravitational lensing by rotating global monopoles has been investigated in Reference [42] and more recently in Reference [43]. Among other things, global monopoles are expected to rotate and to carry magnetic charges.

In this paper, we aim to study the impact of the rotating global monopole black hole surrounded by perfect fluid on the black hole shadow. In Section 2, we consider the gravitational field of a static dyonic black hole (SDBH) with a global monopole surrounded by perfect fluid. In Section 3, by applying a complex coordinate transformation known as the Newman-Janis method [44] we find the spacetimes of a rotating dyonic black hole (RDBH) with a global monopole surrounded by perfect fluid. In Section 4, we consider the null geodesics using Hamilton-Jacobi equation. In Section 5, we study the impact of dark matter, dust and radiation on the shape of global monopole shadow. In Section 6, we study the energy emission rate. Finally in Section 7, we comment on our results.

2. An SDBH with a Global Monopole in Perfect Fluid

The action, $S^{(EM)}$, for Einstein Maxwell gravity along with actions $S^{(D)}$ and $\mathcal{S}$ respectively defining presence of a global monopole and matter distribution, can be altogether written as:

$$S = S^{(EM)} + S^{(D)} + \mathcal{S}. \tag{1}$$

The Einstein-Maxwell action $S^{(EM)}$ is given by:

$$S^{(EM)} = \int \sqrt{-g} d^4x \left(\frac{\mathcal{R}}{2\kappa} - \frac{1}{4} F_{\mu\nu} F^{\mu\nu} \right), \tag{2}$$

where $\kappa = 8\pi$. The quantities g, $\mathcal{R}$ and $F_{\mu\nu}$ are, respectively, the determinant of the metric $g_{\mu\nu}$ associated to the gravitational field, the scalar invariant and the electromagnetic tensor. Also $\mu, \nu = 0, 1, 2, 3$.

The corresponding Einstein field equations read:

$$\mathcal{R}_{\mu\nu} - \frac{1}{2} g_{\mu\nu} \mathcal{R} = 8\pi T_{\mu\nu}. \tag{3}$$

While the corresponding Maxwell equations are:

$$\nabla_\mu F^{\mu\nu} = 0. \tag{4}$$

Here $T_{\mu\nu}$ is the total stress energy tensor which we discuss later in this section. Since we are considering a dyonic black hole, which means that it is comprised of both electric charge Q_E and magnetic charge Q_M, the electromagnetic potential has two non zero terms, that is, [45,46]:

$$\mathbf{A} = \frac{Q_E}{r}dt - Q_M \cos\theta d\varphi. \tag{5}$$

The only non-vanishing components of the electromagnetic tensor:

$$F_{tr} = -F_{rt} = \frac{Q_E}{r^2}, \quad F_{\theta\varphi} = -F_{\varphi\theta} = Q_M \sin\theta. \tag{6}$$

Now the action $S^{(D)}$ corresponds to the matter having a defect– a global monopole which is a heavy object formed in the phase transition of a system composed by a self-coupling scalar triplet field Φ^s, where s runs from 1 to 3. Thus the action in presence of a matter field Φ^s coupled to gravity that characterizes a global monopole [40]:

$$S^{(D)} = \int \sqrt{-g}d^4x \left(\frac{1}{2}g^{\mu\nu}\partial_\mu\Phi^s\partial_\nu\Phi^s - \frac{\lambda}{4}\left(\Phi^2 - \gamma^2\right)^2\right), \tag{7}$$

where $\Phi^2 = \Phi^s\Phi^s$, while λ is the self-interaction term and γ is the scale of a gauge-symmetry breaking. The monopole can be described through the field configuration $\Phi^s = \frac{\gamma h(r)x^s}{|x|}$, in which $x^s = \{r\sin\theta\cos\varphi, r\sin\theta\sin\varphi, r\cos\theta\}$, such that $|x| = r^2$, and $h(r)$ is a function of radial coordinate r. The field equations for the scalar field Φ^s reduces to a single equation for $h(r)$ given as:

$$f(r)h''(r) + \left[\frac{2f(r)}{r} + \frac{1}{2f(r)}(f^2(r))'\right]h'(r) - \frac{2h(r)}{r^2} - \lambda\gamma^2 h(r)\left(h^2(r) - 1\right) = 0. \tag{8}$$

With these equations in mind, and without loss of generality we can choose a spherically symmetric metric written as follows:

$$ds^2 = -f(r)dt^2 + \frac{dr^2}{f(r)} + r^2 d\theta^2 + r^2 \sin^2\theta d\varphi^2. \tag{9}$$

In our case the total stress-energy momentum reads:

$$T_{\mu\nu} = T^{(EM)}_{\mu\nu} + T^{(D)}_{\mu\nu} + \mathcal{T}_{\mu\nu} \tag{10}$$

in which:

$$T^{(EM)}_{\mu\nu} = \frac{1}{4\pi}\left(F_{\mu\sigma}F_\nu{}^\sigma - \frac{1}{4}g_{\mu\nu}F_{\rho\sigma}F^{\rho\sigma}\right), \tag{11}$$

$$T^{(D)}_{\mu\nu} = \partial_\mu\phi^a\partial_\nu\phi^a - \frac{1}{2}g_{\mu\nu}g^{\rho\sigma}\partial_\rho\phi^a\partial_\sigma\phi^a - \frac{g_{\mu\nu}\lambda}{4}\left(\phi^2 - \gamma^2\right)^2, \tag{12}$$

and $\mathcal{T}_{\mu\nu}$ is the energy-momentum tensor of the surrounding matter. The energy momentum-tensor of the surrounding fluid has the following components [47]:

$$\mathcal{T}^t{}_t = \mathcal{T}^r{}_r = -\rho, \tag{13}$$

and:

$$\mathcal{T}^\theta{}_\theta = \mathcal{T}^\varphi{}_\varphi = \frac{1}{2}(1+3\omega)\rho. \tag{14}$$

Outside the core $h \to 1$ and the energy-momentum tensor of the monopole has the following components [40]:

$$T^{(D)t}{}_t = T^{(D)r}{}_r = -\gamma^2 \left[\frac{h^2}{r^2} + f(r)\frac{(h')^2}{2} + \frac{\lambda \gamma^2}{4}(h^2 - 1)^2 \right] \to -\frac{\gamma^2}{r^2}, \quad (15)$$

$$T^{(D)\theta}{}_\theta = T^{(D)\varphi}{}_\varphi = -\gamma^2 \left[f(r)\frac{(h'(r))^2}{2} + \frac{\lambda \gamma^2}{4}(h^2(r) - 1)^2 \right] \to 0. \quad (16)$$

The surrounding matter, whose action is denoted by $\mathcal{S}$ in Equation (1), can generally be a dust, radiation, quintessence, cosmological constant, phantom field or even any combination of them. Thus, the Einstein's field equations yield:

$$\frac{rf'(r) + f(r) - 1}{r^2} + \frac{8\pi\gamma^2}{r^2} + \frac{Q_E^2}{r^4} + \frac{Q_M^2}{r^4} + 8\pi\rho = 0, \quad (17)$$

$$\frac{rf''(r) + 2f'(r)}{2r} - \frac{Q_E^2}{r^4} - \frac{Q_M^2}{r^4} - 4\pi\rho(3\omega + 1) = 0. \quad (18)$$

Now by solving the set of differential equations (18) and (19) one obtains the following general solution for the metric:

$$f(r) = 1 - 8\pi\gamma^2 - \frac{2M}{r} + \frac{Q_E^2}{r^2} + \frac{Q_M^2}{r^2} - \frac{v}{r^{1+3\omega}}, \quad (19)$$

with the energy density in the form:

$$\rho = -\frac{3\omega v}{8\pi r^{3(1+\omega)}}. \quad (20)$$

Note that, v is the perfect fluid parameter. From the weak energy condition it follows the positivity of the energy density of the surrounding field, $\rho \geq 0$, which should satisfy the following constraint $\omega v \leq 0$.

3. An RDBH with a Global Monopole in Perfect Fluid

We now extend the study of static global monopole solution and obtain its rotating counterpart. For this we apply Newman-Janis formalism to the metric (9) along with (19). As a first step to this formalism, we transform Boyer-Lindquist (BL) coordinates (t, r, θ, ϕ) to Eddington-Finkelstein (EF) coordinates (u, r, θ, ϕ). This can be achieved by using the coordinate transformation:

$$dt = du + \frac{dr}{1 - 8\pi\gamma^2 - \frac{2M}{r} + \frac{Q_E^2}{r^2} + \frac{Q_M^2}{r^2} - \frac{v}{r^{3\omega+1}}}, \quad (21)$$

which yields line element in the form:

$$ds^2 = -\left(1 - 8\pi\gamma^2 - \frac{2M}{r} + \frac{Q_E^2}{r^2} + \frac{Q_M^2}{r^2} - \frac{v}{r^{3\omega+1}}\right) du^2 - 2dudr + r^2 d\Omega^2, \quad (22)$$

where $d\Omega^2 = d\theta^2 + \sin^2\theta d\phi^2$. It is worth noting that, compared to the previous work in Reference [41], we shall use the metric form (9) along with $f(r)$ given by Equation (19) to obtain a simple metric for the rotating black hole with a global monopole. This metric can be decomposed in terms of null tetrads as:

$$g^{\mu\nu} = -l^\mu n^\nu - l^\nu n^\mu + m^\mu \overline{m}^\nu + m^\nu \overline{m}^\mu, \quad (23)$$

where the null vectors are defined as:

$$l^\mu = \delta^\mu_r, \tag{24}$$

$$n^\mu = \delta^\mu_u - \frac{1}{2}\left(1 - 8\pi\gamma^2 - \frac{2M}{r} + \frac{Q_E^2}{r^2} + \frac{Q_M^2}{r^2} - \frac{v}{r^{3\omega+1}}\right)\delta^\mu_r, \tag{25}$$

$$m^\mu = \frac{1}{\sqrt{2}r}\left(\delta^\mu_\theta + \frac{i}{\sin\theta}\delta^\mu_\phi\right), \tag{26}$$

$$\bar{m}^\mu = \frac{1}{\sqrt{2}r}\left(\delta^\mu_\theta - \frac{i}{\sin\theta}\delta^\mu_\phi\right). \tag{27}$$

It is obvious from the notation that $\bar{m}^\mu$ is a complex conjugate of m^μ. These vectors further satisfy the conditions for normalization, orthogonality and isotropy as:

$$l^\mu l_\mu = n^\mu n_\mu = m^\mu m_\mu = \bar{m}^\mu \bar{m}_\mu = 0, \tag{28}$$

$$l^\mu m_\mu = l^\mu \bar{m}_\mu = n^\mu m_\mu = n^\mu \bar{m}_\mu = 0, \tag{29}$$

$$-l^\mu n_\mu = m^\mu \bar{m}_\mu = 1. \tag{30}$$

Following the Newman–Janis prescription we write:

$$x'^\mu = x^\mu + ia(\delta^\mu_r - \delta^\mu_u)\cos\theta \rightarrow \begin{cases} u' = u - ia\cos\theta, \\ r' = r + ia\cos\theta, \\ \theta' = \theta, \\ \phi' = \phi. \end{cases} \tag{31}$$

in which a stands for the rotation parameter. Next, let the null tetrad vectors $Z^a = (l^a, n^a, m^a, \bar{m}^a)$ undergo a transformation given by $Z^\mu = (\partial x^\mu/\partial x'^\nu)Z'^\nu$, following:

$$l'^\mu = \delta^\mu_r, \tag{32}$$

$$n'^\mu = \delta^\mu_u - \frac{1}{2}\mathcal{F}\delta^\mu_r, \tag{33}$$

$$m'^\mu = \frac{1}{\sqrt{2\Sigma}}\left[(\delta^\mu_u - \delta^\mu_r)ia\sin\theta + \delta^\mu_\theta + \frac{i}{\sin\theta}\delta^\mu_\phi\right], \tag{34}$$

$$\bar{m}'^\mu = \frac{1}{\sqrt{2\Sigma}}\left[(\delta^\mu_u - \delta^\mu_r)ia\sin\theta + \delta^\mu_\theta + \frac{i}{\sin\theta}\delta^\mu_\phi\right], \tag{35}$$

where we replaced $f(r)$ to $\mathcal{F}(r,a,\theta)$ and $h(r) = r^2$ to $\Sigma(r,a,\theta)$. With the help of the above equations the contravariant components of new metric are computed as:

$$g^{uu} = \frac{a^2\sin^2\theta}{\Sigma}, \quad g^{u\phi} = \frac{a}{\Sigma}, \quad g^{ur} = 1 - \frac{a^2\sin^2\theta}{\Sigma},$$

$$g^{rr} = \mathcal{F} + \frac{a^2\sin^2\theta}{\Sigma}, \quad g^{r\phi} = -\frac{a}{\Sigma}, \quad g^{\theta\theta} = \frac{1}{\Sigma},$$

$$g^{\phi\phi} = \frac{1}{\Sigma\sin^2\theta}. \tag{36}$$

The new metric is found as follows:

$$ds^2 = -\mathcal{F}du^2 - 2dudr + 2a\sin^2\theta(\mathcal{F}-1)dud\phi + 2a\sin^2\theta\, drd\phi + \Sigma d\theta^2$$
$$+ \sin^2\theta\left[\Sigma + a^2(2-\mathcal{F})\sin^2\theta\right]d\phi^2. \tag{37}$$

Using the method without a complexification introduced in Reference [48] we revert the EF coordinates back to BL coordinates by using the following transformation:

$$du = dt + \lambda(r)dr, \quad d\phi = d\varphi + \chi(r)dr, \tag{38}$$

where

$$\lambda(r) = \frac{-a^2 - k(r)}{f(r)h(r) + a^2}, \quad \chi(r) = \frac{-a}{f(r)h(r) + a^2}, \quad k(r) = h(r) = r^2, \tag{39}$$

with

$$\mathcal{F} = \frac{f(r)h(r) + a^2 \cos^2 \theta}{(k(r) + a^2 \cos^2 \theta)^2} \Sigma. \tag{40}$$

Hence the rotating black hole solution in Boyer-Lindquist coordinates turns out to be [49]:

$$ds^2 = -\frac{f(r)h(r) + a^2 \cos^2 \theta}{(k(r) + a^2 \cos^2 \theta)^2} \Sigma dt^2 + 2a \sin^2 \theta \frac{f(r)h(r) - k(r)}{(k(r) + a^2 \cos^2 \theta)^2} \Sigma dt d\varphi + \frac{\Sigma}{f(r)h(r) + a^2} dr^2 + \Sigma d\theta^2$$
$$+ \Sigma \sin^2 \theta \left[1 + a^2 \sin^2 \theta \frac{2k(r) - f(r)h(r) + a^2 \cos^2 \theta}{(k(r) + a^2 \cos^2 \theta)^2} \right] d\varphi^2.$$

Following Reference [48] using the condition $k(r) = h(r) = r^2$ one can find $\Sigma = r^2 + a^2 \cos^2 \theta$. Finally the metric can also be written as:

$$ds^2 = -\left(1 - \frac{r^2(1 - f(r))}{\Sigma}\right) dt^2 - 2a \sin^2 \theta \left(\frac{r^2(1 - f(r))}{\Sigma}\right) dt d\varphi + \frac{\Sigma}{\Delta} dr^2 + \Sigma d\theta^2$$
$$+ \sin^2 \theta \left[\frac{(r^2 + a^2)^2 - a^2 \Delta \sin^2 \theta}{\Sigma}\right] d\varphi^2, \tag{41}$$

where in order to simplify the notation we introduce the following quantities:

$$\Delta = r^2 f(r) + a^2 = r^2 + a^2 - 2Mr - 8\pi r^2 \gamma^2 + Q_E^2 + Q_M^2 - \frac{v}{r^{3\omega - 1}}, \tag{42}$$

where $f(r)$ is given by Equation (19). In this work, we consider three different cases of $\omega = -1/3$ dark matter dominant, 0 (dust dominant) and 1/3 (radiation dominant). For spin $a = 0$, perfect fluid parameter $v = 0$ and no charges, the above metric reduces to Schwarzschild black hole with global monopole [50]. It would certainly be interesting to generalize our solution by including the cosmological constant as well. As a particular example, we point out the Kerr-Newman-NUT black hole which is a subclass of the Plebanski-Demianski class [51], but our solution it seems not to be the case. The shadows of Kerr-Newman-NUT black holes with cosmological constant has been investigated by Grenzebach et al. [52]. The electromagnetic field of a black hole is defined by its vector potential. As already mentioned, in case of a static black hole the vector potential is given by Equation (5). For the rotating case, the Newman-Janis method can also be applied on Equation (5) using a gauge transformation such that $g_{rr} = 0$ and $A_r = 0$. For the detailed procedure the authors refer the readers to Reference [46]. The vector potential computed through Newman-Janis formalism for a rotating dyonic black hole is given by [46]:

$$\mathbf{A} = \left(\frac{rQ_E - aQ_M \cos \theta}{\Sigma}\right) dt + \left(-\frac{ra}{\Sigma} Q_E \sin^2 \theta + \frac{r^2 + a^2}{\Sigma} Q_M \cos \theta\right) d\varphi. \tag{43}$$

It has been shown in Reference [48] that metric similar to (41) satisfies the Einstein field equations. For the Einstein tensor $G_{\mu\nu}$ and energy-momentum tensor $T_{\mu\nu}$, the Einstein field equations are given by $G_{\mu\nu} = R_{\mu\nu} - 1/2 g_{\mu\nu} \mathcal{R} = 8\pi T_{\mu\nu}$. For simplicity, let $f(r) = 1 - 2F(r)/r^2$, where $F(r) = 4\pi\gamma^2 r^2 + Mr - (Q_E^2 + Q_M^2)/2 + v r^{1-3\omega}/2$, then the nonvanishing components of $G_{\mu\nu}$ are:

$$
\begin{aligned}
G_{tt} &= \frac{2}{\Sigma^3}\left(2F(r) - \left((r^2 + a^2) + a^2\sin^2\theta\right)\right)(F(r) - rF'(r)) - \frac{a^2 \sin^2\theta}{\Sigma^2}F''(r), \\
G_{rr} &= \frac{2}{\Sigma\Delta}(F(r) - rF'(r)), \\
G_{\theta\theta} &= \frac{-2}{\Sigma}(F(r) - rF'(r)) - F''(r), \\
G_{t\varphi} &= \frac{4a\sin^2\theta}{\Sigma^3}\left((r^2 + a^2) - F(r)\right)(F(r) - rF'(r)) + \frac{a}{\Sigma^2}(r^2 + a^2)\sin^2\theta F''(r), \\
G_{\varphi\varphi} &= \frac{\sin^2\theta}{\Sigma^3}\left(4a^2\sin^2\theta F(r) - (r^2 + a^2)\left(2(r^2 + a^2) + a^2\sin^2\theta\right)\right)(F(r) - rF'(r)) - \frac{(r^2 + a^2)\sin^2\theta}{\Sigma^2}F''(r).
\end{aligned}
\qquad (44)
$$

In terms of the orthogonal basis, for the metric (41):

$$
e_t^\mu = \frac{1}{\sqrt{\Sigma\Delta}}\left(r^2 + a^2, 0, 0, a\right), \quad e_r^\mu = \frac{1}{\sqrt{\Sigma}}(0, 1, 0, 0),
\qquad (45)
$$
$$
e_\theta^\mu = \frac{1}{\sqrt{\Sigma}}(0, 0, 1, 0), \quad e_\varphi^\mu = \frac{1}{\sqrt{\Sigma\Delta}}\left(a\sin^2\theta, 0, 0, 1\right).
$$

and the Einstein tensor $G_{\mu\nu}$, the energy momentum tensor is expressed as:

$$
\begin{aligned}
p_t &= \frac{1}{8\pi}e_t^\mu e_t^\nu G_{\mu\nu}, \quad p_r = \frac{1}{8\pi}e_r^\mu e_r^\nu G_{\mu\nu}, \\
p_\theta &= \frac{1}{8\pi}e_\theta^\mu e_\theta^\nu G_{\mu\nu}, \quad p_\varphi = \frac{1}{8\pi}e_\varphi^\mu e_\varphi^\nu G_{\mu\nu}.
\end{aligned}
\qquad (46)
$$

Equations (41)–(46) gives the components for energy momentum tensor as:

$$
\begin{aligned}
p_t &= \frac{1}{8\pi\Sigma^2}\left(8\pi\gamma^2 r^2 - 3v\omega r^{1-3\omega} + (Q_E^2 + Q_M^2)\right) = -p_r, \\
p_\theta &= -p_r - \frac{1}{8\pi\Sigma}\left(8\pi\gamma^2 - \frac{3v\omega(1 - 3\omega)}{2r^{1+3\omega}}\right) = p_\varphi.
\end{aligned}
\qquad (47)
$$

Analogous to a Kerr black hole, a ring singularity resides inside the black hole defined by metric (41). This can be demonstrated by computing the points at which the Kretschmann scalar $\mathcal{K}_s = R_{\mu\nu\sigma\rho}R^{\mu\nu\sigma\rho}$ turns to infinity. For the metric (41), the Kretschmann scalar has the value:

$$
\mathcal{K}_s = \frac{Z(r, a, \theta, Q_E, Q_M, \omega, v, \gamma)}{r^{2(1+3\omega)}(r^2 + a^2 \cos^2\theta)^2}.
\qquad (48)
$$

where $Z(r, a, \theta, Q_E, Q_M, \omega, v, \gamma)$ is a tedious function. From the above expression, we observe that for $\omega = -1/3, 0, 1/3$ the poles lie $r^2 + a\cos^2\theta = 0$ or when $r = 0$ and $\theta = \pi/2$. This leads us to the interpretation that a test particle moving in an equatorial plane $\theta = \pi/2$ will hit the singularity at $r = 0$.

3.1. Surface Topology

It is interesting to determine the surface topology of the global monopole spacetime at the event horizon. At a fixed moment in time t, and a constant $r = r_+$, the metric (41) reduces to:

$$
ds^2 = \Sigma(r_+, \theta)d\theta^2 + \left(2Mr_+ + 8\pi r_+^2 \gamma^2 - Q_E^2 - Q_M^2 + \frac{v}{r_+^{3\omega-1}}\right)^2 \frac{\sin^2\theta}{\Sigma(r_+, \theta)}d\varphi^2,
\qquad (49)
$$

The above metric has the following determinant:

$$\det g^{(2)} = \left(2Mr_+ + 8\pi r_+^2 \gamma^2 - Q_E^2 - Q_M^2 + \frac{v}{r_+^{3\omega-1}}\right)^2 \sin^2\theta. \quad (50)$$

Theorem 1. *Let $\mathcal{M}$ be a compact orientable surface with metric $g^{(2)}$, and let K be the Gaussian curvature with respect to $g^{(2)}$ on $\mathcal{M}$. Then, the Gauss-Bonnet theorem states that:*

$$\iint_{\mathcal{M}} K \, dA = 2\pi \chi(\mathcal{M}). \quad (51)$$

Note that dA is the surface line element of the 2-dimensional surface and $\chi(\mathcal{M})$ is the Euler characteristic number. It is convenient to sometimes express the above theorem in terms of the Ricci scalar, in particular for the 2-dimensional surface there is a simple relation between the Gaussian curvature and Ricci scalar given by:

$$K = \frac{\mathcal{R}}{2}. \quad (52)$$

Yielding the following from:

$$\frac{1}{4\pi} \iint_{\mathcal{M}} \mathcal{R} \, dA = \chi(\mathcal{M}). \quad (53)$$

A straightforward calculation using the metric (49) yields the following result for the Ricci scalar:

$$\mathcal{R} = \frac{2(r_+^2 + a^2)(r_+^2 - 3a^2 \cos^2\theta)}{(r_+^2 + a^2 \cos^2\theta)^3} \quad (54)$$

From the GBT we find:

$$\chi(\mathcal{M}) = \frac{1}{4\pi} \int_0^{2\pi} \int_0^{\pi} \left[\frac{2(r_+^2 + a^2)(r_+^2 - 3a^2 \cos^2\theta)}{(r_+^2 + a^2 \cos^2\theta)^3}\right] \sqrt{\det g^{(2)}} d\theta d\varphi. \quad (55)$$

Finally, solving the integral we find:

$$\chi(\mathcal{M}) = 2. \quad (56)$$

Hence the surface topology is a 2-sphere at the event horizon, since we know that $\chi(\mathcal{M})_{sphere} = 2$.

3.2. Shape of Ergoregion

Let us now proceed to study the shape of the ergoregion of an RDBH with a global monopole. In particular we shall be interested to plot the shape of the ergoregion in the xz-plane. Recall that the horizons of the RDG can be found by solving $\Delta = 0$:

$$r^2 + a^2 - 2Mr - 8\pi r^2 \gamma^2 + Q_E^2 + Q_M^2 - \frac{v}{r^{3\omega-1}} = 0, \quad (57)$$

on the other hand, the limit surfaces or inner and outer ergosurface is given by $g_{tt} = 0$, i.e.,

$$r^2 + a^2 \cos^2\theta - 2Mr - 8\pi r^2 \gamma^2 + Q_E^2 + Q_M^2 - \frac{v}{r^{3\omega-1}} = 0. \quad (58)$$

There is an interesting process which relies on the presence of an ergoregion, namely from such a rotating black hole energy can be extracted, and is known as the Penrose process. In Figure 1 we plot the shape of ergoregion for different values of a, ω, γ, and v. One can observe that the event horizon and static limit surface meet at the poles while the region between them is the ergoregion which supports negative energy orbits. Furthermore, the shape of ergoregion depends on the spin a, however due to the small values of v we observe small changes related to the value of ω.

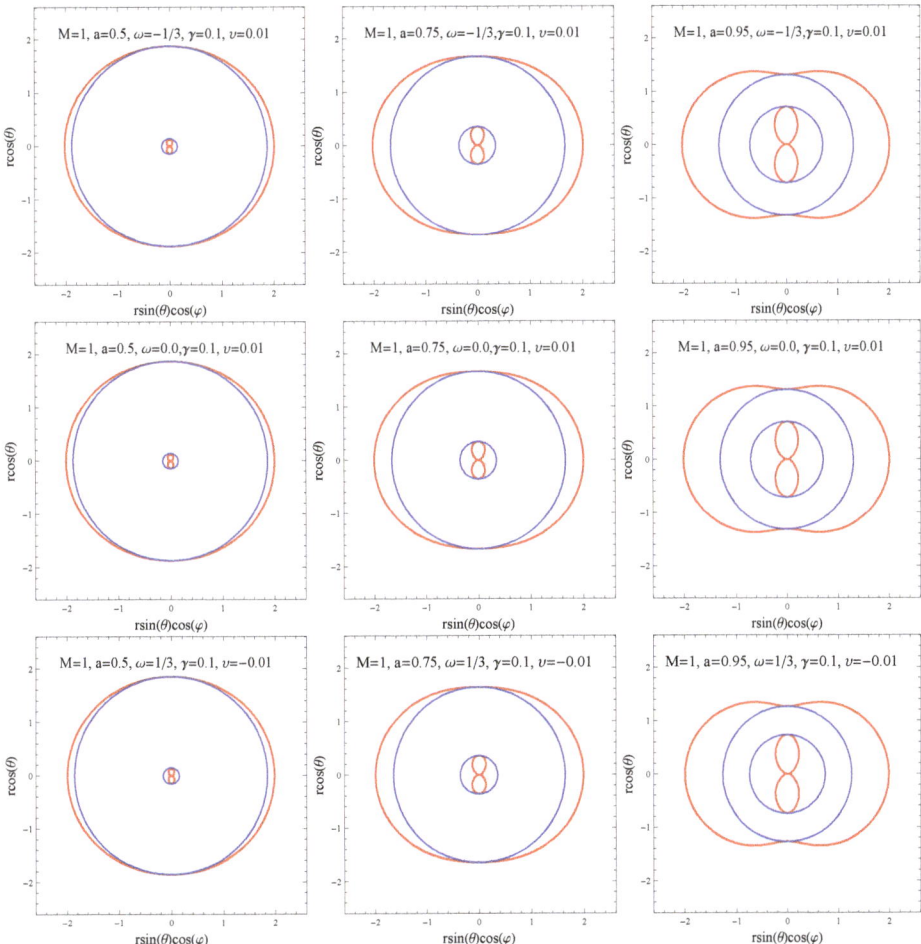

Figure 1. Plots showing the shape of ergoregion in xz-plane for different values of a, ω, and v. We have chosen $Q_E = Q_M = 0.1$ in all plots. The blue and the red lines correspond to horizons and static limit surfaces, respectively. The outer red line corresponds to the static limit surface, whereas the two blue lines correspond to the two horizons. Due to the small values of v and a arbitrary value of a we observe almost indistinguishable plots for the shape of the ergoregion.

4. Null Geodesics

Our main objective is to study the shadow cast by the black hole defined by metric (41). To do so, we first need to analyze the geodesics structure of photons moving around the compact gravitational source. This will enable us to detect the unstable photon orbits which in turn defines the boundary of the shadow.

To observe the null geodesics around the RDGM present in perfect fluid, we consider the Hamilton-Jacobi method. The Hamilton-Jacobi equation is given by:

$$\partial_\tau \mathcal{J} = -\mathcal{H}. \tag{59}$$

In the above equation:

On Left Side $\mathcal{J}$ is the Jacobi action, defined as the function of affine parameter τ and coordinates x^μ
i.e., $\mathcal{J} = \mathcal{J}(\tau, x^\mu)$.
On Right Side $\mathcal{H}$ is the Hamiltonian of test particle's motion and is equivalent to $g^{\mu\nu}\partial_\mu \mathcal{J}\, \partial_\nu \mathcal{J}$.

In the spacetime under consideration, along the photon geodesics the energy E and momentum L, defined respectively by Killing fields $\xi_t = \partial_t$ and $\xi_\phi = \partial_\phi$, are conserved. The mass $m = 0$ of the photon is also constant. Using these constants of motion we can thus separate the Jacobi function as:

$$\mathcal{J} = \frac{1}{2}m^2\tau - Et + L\phi + \mathcal{J}_r(r) + \mathcal{J}_\theta(\theta), \tag{60}$$

where the functions $\mathcal{J}_r(r)$ and $\mathcal{J}_\theta(\theta)$ respectively depends on coordinates r and θ.

Combining Equations (59) and (60) yields a set of equations, which describes the dynamics of a test particle around the rotating black hole in perfect fluid matter, as:

$$\Sigma \frac{dt}{d\tau} = \frac{r^2 + a^2}{\Delta}[E(r^2 + a^2) - aL] - a(aE\sin^2\theta - L), \tag{61}$$

$$\Sigma \frac{dr}{d\tau} = \sqrt{\mathcal{R}(r)}, \tag{62}$$

$$\Sigma \frac{d\theta}{d\tau} = \sqrt{\Theta(\theta)}, \tag{63}$$

$$\Sigma \frac{d\varphi}{d\tau} = \frac{a}{\Delta}[E(r^2 + a^2) - aL] - \left(aE - \frac{L}{\sin^2\theta}\right), \tag{64}$$

where $\mathcal{R}(r)$ and $\Theta(\theta)$ read as:

$$\mathcal{R}(r) = [E(r^2 + a^2) - aL]^2 - \Delta[m^2 r^2 + (aE - L)^2 + \mathcal{K}], \tag{65}$$

$$\Theta(\theta) = \mathcal{K} - \left(\frac{L^2}{\sin^2\theta} - a^2 E^2\right)\cos^2\theta, \tag{66}$$

with $\mathcal{K}$ the Carter constant.

5. Circular Orbits

Now we consider a gravitational source placed between a light emitting source and an observer at infinity. The photons emitted from the light source will form two kinds of trajectories—the ones which eventually fall into the black hole and the ones which scatter away from it. The region separating these trajectories contains unstable circular orbits. These unstable circular orbits form a dark region in sky thus forming the contour of the shadow. In this section we intend to discuss the presence of unstable circular orbits around the black hole under consideration. For this we consider photon as a test particle and hence take $m = 0$. We can express the radial geodesic equation in terms of effective potential V_{eff} of photon's radial motion as:

$$\Sigma^2 \left(\frac{dr}{d\tau}\right)^2 + V_{\text{eff}} = 0.$$

For our convenience we introduce two independent parameters ξ and η [53] as:

$$\xi = L/E, \quad \eta = \mathcal{K}/E^2. \tag{67}$$

The effective potential in terms of these two parameters is then expressed as:

$$V_{\text{eff}} = \Delta((a - \xi)^2 + \eta) - (r^2 + a^2 - a\,\xi)^2, \tag{68}$$

where we have replaced V_{eff}/E^2 by V_{eff}. Figure 2 shows the variation in effective potential associated with the radial motion of photons. From the figure we observe that in all three cases the value of effective potential decreases with increase in parameter γ. Now the circular photon orbits exists when at some constant $r = r_c$ the conditions:

$$V_{\text{eff}}(r) = 0, \qquad \frac{dV_{\text{eff}}(r)}{dr} = 0 \tag{69}$$

are satisfied. We then use Equation (68) in Equation (69) and thus obtain:

$$[\eta + (\xi - a)^2]\Delta - (r^2 + a^2 - a\xi)^2 = 0, \tag{70}$$
$$4(r^2 + a^2 - a\xi) - [\eta + (\xi - a)^2]\mathcal{A}(r) = 0, \tag{71}$$

where:

$$\mathcal{A}(r) = 1 - 8\pi\gamma^2 - \frac{M}{r} + \left(\frac{3\omega - 1}{2}\right)\frac{v}{r^{1+3\omega}}. \tag{72}$$

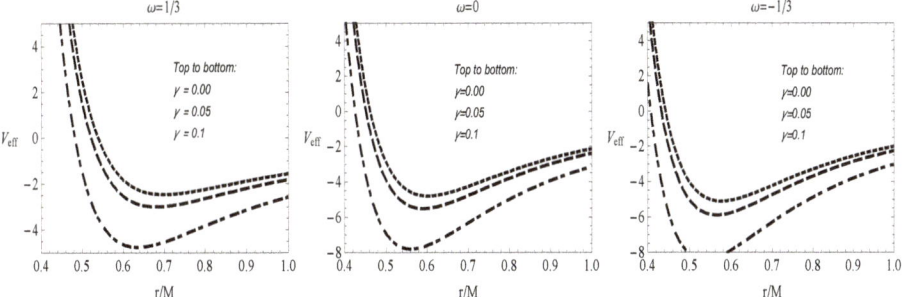

Figure 2. The effective potential of photon moving in equatorial plane, with respect to its radial motion: $\omega = 1/3$ for radiation, $\omega = 0$ for dust and $\omega = -1/3$ for dark matter.

Combining Equations (70) and (71) results in:

$$a\,\xi = r^2 + a^2 - \frac{2\,\Delta}{\mathcal{A}(r)}, \tag{73}$$

$$\eta = \frac{4\Delta}{\mathcal{A}(r)^2} - \frac{1}{a^2}\left(r^2 - \frac{\Delta}{\mathcal{A}(r)^2}\right)^2 \tag{74}$$

It is worth mentioning here that impact parameters, ξ and η, will be affected not just by radial coordinate r, spin parameter a and mass of black hole M but also by electric charge Q_E, magnetic charge Q_M, monopole parameter γ and perfect fluid parameter v. The unstable circular orbits are located at local maxima of the potential curves, that is, when $V''_{\text{eff}} < 0$ or:

$$\left(\Delta'^2 + 2\Delta\Delta''\right)r + 2\Delta\Delta' > 0 \tag{75}$$

6. Silhoutte of Black Holes

In this section, we extend our calculations to observe shadow of RDGM surrounded by perfect fluid. To gain the optical image we specify the observer at position (r_o, θ_o), where $r_o = r \to \infty$ and θ_o is the angular coordinate at infinity, on observer's sky. The new coordinates, also widely known as celestial coordinates, α and β, are then introduced. These coordinates are selected such that α and β correspond to the apparent perpendicular distance of the image from axis of symmetry and its projection on the equatorial plane, respectively.

Due to the presence of global monopole, we have asymptotically non flat solutions due to the global nontrivial topology. Now we obtain the proper celestial coordinates for the asymptotically non-flat solution by adopting [10]:

$$\alpha = \lim_{r \to \infty} -r \frac{p^{(\phi)}}{p^{(t)}} \tag{76}$$

$$\beta = \lim_{r \to \infty} r \frac{p^{(\theta)}}{p^{(t)}} \tag{77}$$

where $(p^{(t)}, p^{(r)}, p^{(\theta)}, p^{(\phi)})$ are the tetrad components of the photon momentum with respect to locally nonrotating reference frame. So basically one can define the observer's sky as the usual cases in which the observer bases $e^{\mu}_{(\nu)}$ can be expanded as a form in the coordinate bases. In the limit $r \to \infty$ can relate the above coordinates to parameters ξ and η, which then yield:

$$\alpha = -\sqrt{1 - 8\pi\gamma^2} \frac{\xi}{\sin\theta} \tag{78}$$

$$\beta = \pm\sqrt{1 - 8\pi\gamma^2} \sqrt{\eta + a^2 \cos^2\theta - \xi^2 \cot^2\theta}, \tag{79}$$

for the case $\omega = 0$ and $\omega = 1/3$. And similarly:

$$\alpha = -\sqrt{1 - 8\pi\gamma^2 - v} \frac{\xi}{\sin\theta} \tag{80}$$

$$\beta = \pm\sqrt{1 - 8\pi\gamma^2 - v} \sqrt{\eta + a^2 \cos^2\theta - \xi^2 \cot^2\theta} \tag{81}$$

for the case of quintessence, that is, $\omega = -1/3$. We observe that in the dark matter case there is a similar contribution term compared to the global monopole. In the limit $\gamma \to 0$ and $v \to 0$ we obtain the usual relations for celestial coordinates for the asymptotically flat solution. We expect that the parameters involved in RDGM in the presence of a perfect fluid will affect the shape of its shadow. This can be clearly confirmed through Equation (81) as it depends not only on spin parameter a and angular coordinate θ_o but also on γ, ω and perfect fluid parameter v. Later, we will justify our results also through graphical interpretations.

As our observer is placed in the equatorial plane ($\theta = \pi/2$), α and β reduce to:

$$\alpha = -\sqrt{1 - 8\pi\gamma^2} \, \xi \tag{82}$$

$$\beta = \pm\sqrt{1 - 8\pi\gamma^2} \, \sqrt{\eta}, \tag{83}$$

for the case $\omega = 0$ and $\omega = 1/3$. And

$$\alpha = -\sqrt{1 - 8\pi\gamma^2 - v} \, \xi \tag{84}$$

$$\beta = \pm\sqrt{1 - 8\pi\gamma^2 - v} \, \sqrt{\eta} \tag{85}$$

for the case $\omega = -1/3$. Figures 3 and 4 show deformation in the shapes of the shadow with respect to monopole parameter γ and perfect fluid parameter v, respectively. It is a well known observation now that the rotational effect in a black hole distorts its shape. That being said, we notice in Figure 3 that for small spin parameter, a, the shadow of the black hole maintains a circular shape along with the increase in its size with the inclination of γ. As for larger spin values, the shadow is clearly distorted and matches with its Kerr counterpart in perfect fluid [19] for $\gamma = 0$. Figure 4 shows the effect of parameter v on the rotating dyonic black hole with a global monopole present in perfect fluid. It is noticed in Figure 4 that as perfect fluid parameter, v, increases the size of the shadow also increases. A distortion is noticed in shape of the shadow when the spin parameter a is increased. Also, in the case of dark matter and dust, there is a significant change in the size of the shadow with respect to v.

On the other hand, in the case of radiation we do not observe any significant effect of perfect fluid parameter v, in fact the effect is negligibly small.

In [10], the authors introduce two observables, radius R_s and distortion δ_s, to analyze the size and form of the shadow. The first observable R_s is the approximate radius of the shadow. It is defined by considering a reference circle passing through three points on the boundary of the shadow, such that $(\alpha_{tp}, \beta_{tp})$ is the top most point on the shadow, $(\alpha_{bm}, \beta_{bm})$ is the bottom most point on the shadow and $(\alpha_r, 0)$ is the point corresponding to unstable circular orbit seen by an observer on reference frame. Thus:

$$R_s = \frac{(\alpha_{tp} - \alpha_r)^2 + \beta_{tp}^2}{2|\alpha_{tp} - \alpha_r|}. \tag{86}$$

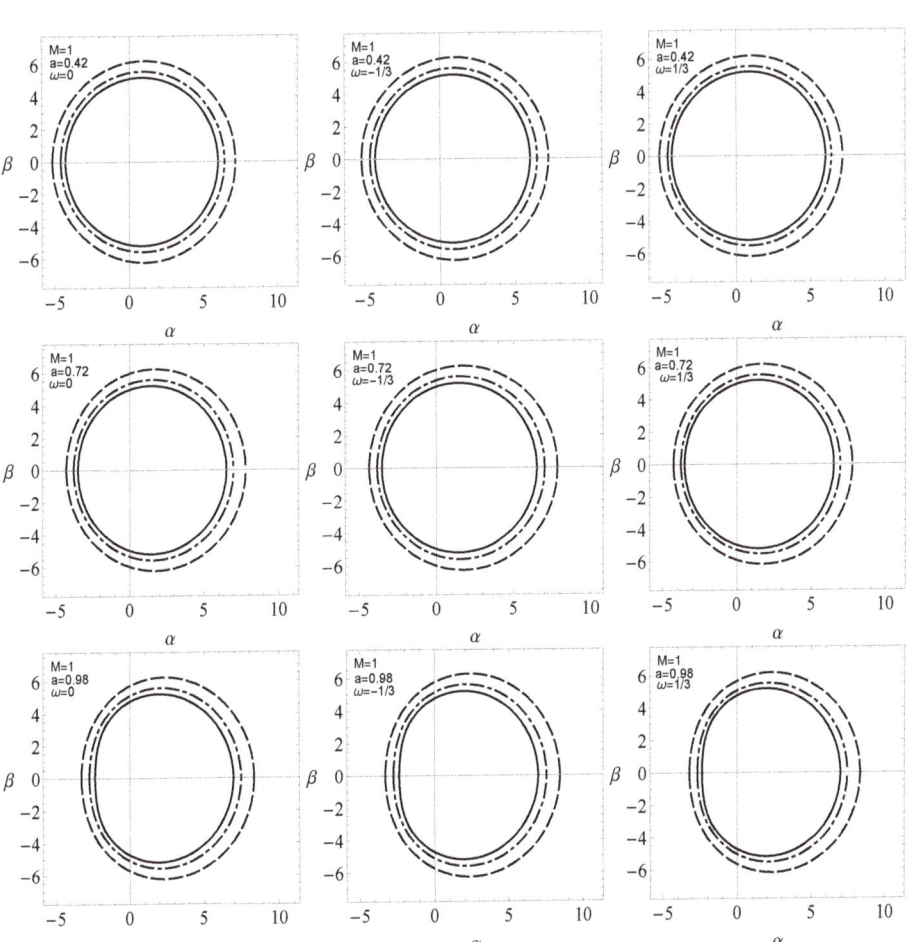

Figure 3. Variation in shape of a rotating dyonic global monopole surrounded by a perfect fluid. Magnetic and electric charges are kept constant such that $Q_E = 10^{-2} = Q_M$. In each graph the Kerr case, that is, $\gamma = 0$ and $v = 0$, is represented by a solid line, $\gamma = 0.05$ by dotdashed and $\gamma = 0.08$ by dashed lines. For dark matter ($\omega = -1/3$) and dust ($\omega = 0$) case $v = 0.01$, whereas $v = -0.01$ in the case of radiation ($\omega = 1/3$).

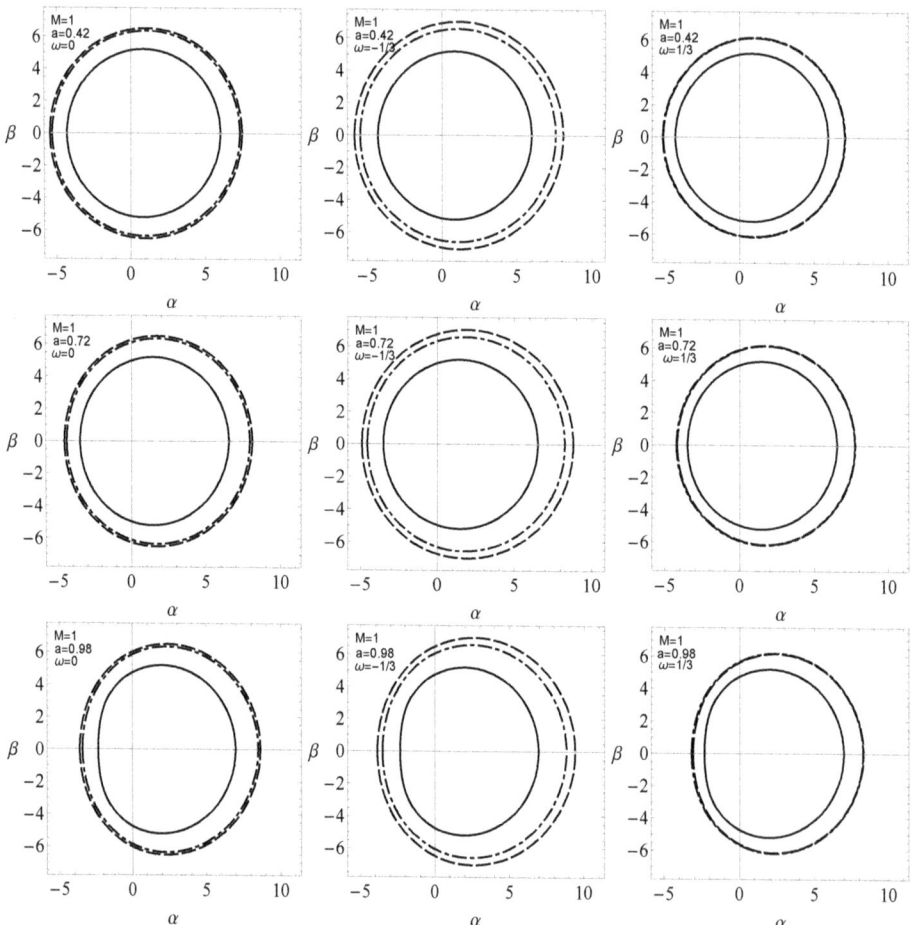

Figure 4. Variation in the shape of a rotating dyonic black hole with global monopole surrounded by a perfect fluid, for different values of perfect fluid parameter v. Magnetic and electric charges along with the global monopole parameter are kept constant such that $Q_E = 10^{-2} = Q_M$ and $\gamma = 0.08$. For dark matter and dust case $v = 0$ (Solid), 0.05 (DotDashed) and 0.1 (Dashed). In the case of radiation $v = 0$ (Solid), -0.01 (DotDashed) and -0.05 (Dashed).

The second observable δ_s is the distortion parameter. Let D_{CS} be the difference between the contour of shadow and the reference circle. Then for the point $(\tilde{\alpha}_p, 0)$ lying on the reference circle and the point $(\alpha_p, 0)$ lying on the contour of the shadow, $D_{CS} = |\tilde{\alpha}_p - \alpha_p|$. Thus:

$$\delta_s = \frac{\tilde{\alpha}_p - \alpha_p}{R_s}.$$

For our case, we consider the points $(\tilde{\alpha}_p, 0)$ and $(\alpha_p, 0)$ to be on the equatorial plane, opposite to the point $(\alpha_r, 0)$. The variations in these observables with respect to monopole parameter γ are graphically presented in Figure 5. The dependence of R_s on parameter γ is such that as γ increases the radius R_s also increases. Thus the size of the shadow increases with increase in monopole parameter γ. Whereas the distortion δ_s decreases monotonically with an increase in γ. This tells us that with respect to circumference of reference circle, the shadow of the rotating black hole is significantly distorted for $\gamma \in [0, 0.1]$ but for $\gamma > 0$ it may not show any distortion and thus we may obtain a perfect circle.

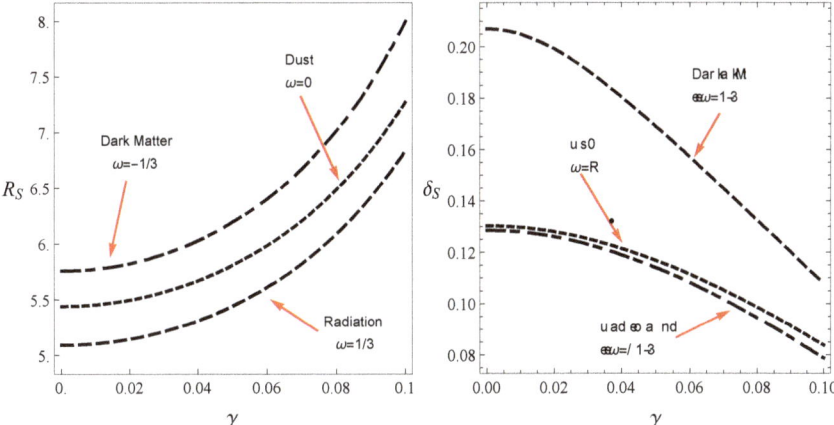

Figure 5. The quantities R_s and δ_s with respect to parameter γ.

As we have considered our observer to be at infinity so in this case the area of the black hole shadow will be approximately equal to high energy absorption cross section as discussed in Reference [18]. For a spherically symmetric black hole the absorption cross section oscillates around Π_{ilm}, a limiting constant value. For a black hole shadow with radius R_s, we adopt the value of Π_{ilm} as calculated by [18]:

$$\Pi_{ilm} \cong \pi R_s^2.$$

The energy emission rate of the black hole is thus defined by:

$$\frac{d^2 E(\sigma)}{d\sigma dt} = 2\pi^2 \frac{\Pi_{ilm}}{e^{\sigma/T} - 1} \sigma^3,$$

where σ is the frequency of the photon and T represents the temperature of the black hole at outer horizon, that is, r_+, given by:

$$\begin{aligned}T(r_+) &= \lim_{r \to r_+} \frac{\partial_r \sqrt{g_{tt}}}{2\pi \sqrt{g_{rr}}} \\ &= \left(2a^2 (f(r) - 1) + r(r^2 + a^2) f'(r)\right) \frac{r}{4\pi (r^2 + a^2)^2}\end{aligned}$$

For all three cases, radiation, dust and dark matter, the energy emission rate is graphically presented in Figure 6 where we notice that the energy emission rate decreases with an increase in parameter γ. A slight shift to the lower frequency is also observed while γ increases. The spin parameter a also effects the shape of the energy emission rate as an abrupt decrease in energy emission rate is noticed for higher spin value.

Recent studies have pointed out a connection between the topology of the shadow shape by introducing a new quantity such as the local curvature radius of the shadow. In our paper, we have applied the Gauss-Bonnet theorem to the horizon surface area to prove that the topology is a 2 sphere. It will be interesting to see if one can find the shadow radius by means of Gauss-Bonnet theorem applied directly to metric by using a relation between the horizon radius and the photon sphere. We are planning to work on such a project in the near future.

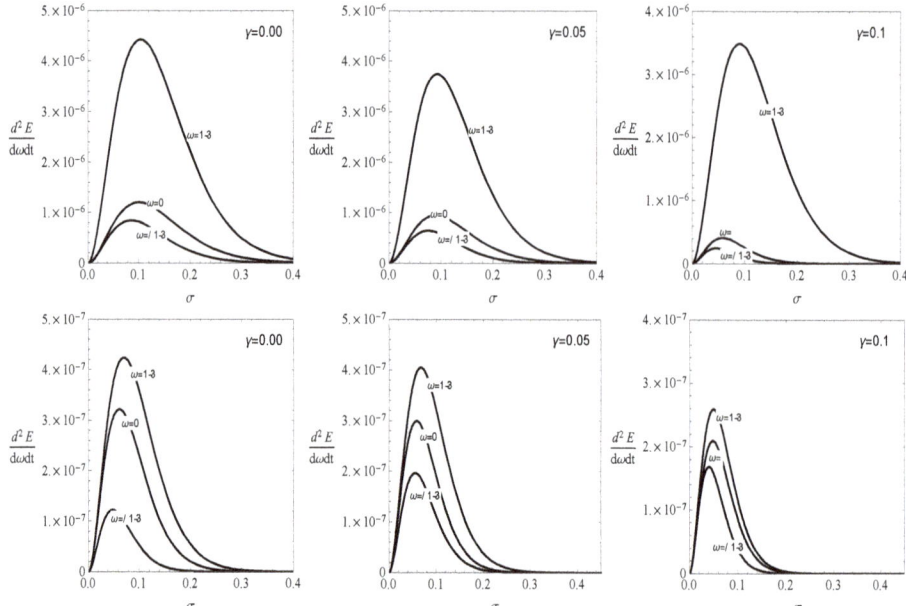

Figure 6. The figure shows the energy emission rate when $a = 0.46$ (**upper panel**) and $a = 0.92$ (**lower panel**).

7. Conclusions

In this paper, we have used the complex transformations pointed out by Newman and Janis to obtain an RDGM solution in the presence of a perfect fluid matter. Using the Gauss-Bonnet theorem we have shown that the surface topology of an RDGM is indeed a 2-sphere. Furthermore by choosing $\omega = -1/3, 0, 1/3$ we have explored the impact of dark matter, dust, radiation, as well as the global monopole parameter γ, and perfect fluid parameters v, on the silhouette of a black hole. We have found that a rotating dyonic black hole with a global monopole retains a circular shape for small spin parameter. Whereas for high spin like $a = 0.98M$ the shadow of RDGM is distorted. Also as monopole parameter γ increases, a slight shift towards the right is also noticed in the shape of the shadow of the black hole under consideration. The two observables, R_s and δ_s, are also discussed. Finally, we analyze the energy emission rate of a rotating dyonic global monopole surrounded by perfect fluid with respect to parameters.

Author Contributions: All three authors contributed equally to the conceptualization, methodology, software, validation, formal analysis, investigation, resources, data curation, writing–original draft preparation, writing–review and editing, visualization, supervision, project administration, funding acquisition. All authors have read and agreed to the published version of the manuscript.

Funding: This research received no external funding.

Acknowledgments: We would like to thank the editor and Marcus Werner for helpful comments and suggestions.

Conflicts of Interest: The authors declare no conflict of interest.

References

1. Broderick, A.; Loeb, A.; Narayan, R. The Event Horizon of Sagittarius A*. *Astrophys. J.* **2009**, *701*, 1357. [CrossRef]
2. Broderick, A.; Narayan, R.; Kormendy, J.; Perlman, E.; Rieke, M.; Doeleman, S. The Event Horizon of M87. *Astrophys. J.* **2015**, *805*, 179. [CrossRef]

3. Cunha, P.V.P.; Herdeiro, C.A.R. Shadows and strong gravitational lensing: A brief review. *Gen. Relat. Gravit.* **2018**, *50*, 42. [CrossRef]
4. Akiyama, K. et al. [Event Horizon Telescope Collaboration]. First M87 Event Horizon Telescope Results. I. The Shadow of the Supermassive Black Hole. *Astrophys. J.* **2019**, *875*, L1.
5. Akiyama, K. et al. [Event Horizon Telescope Collaboration]. First M87 Event Horizon Telescope Results. VI. The Shadow and Mass of the Central Black Hole. *Astrophys. J.* **2019**, *875*, L6.
6. Synge, J.L. The escape of photons from gravitationally intense stars. *Mon. R. Astron. Soc.* **1966**, *131*, 463. [CrossRef]
7. Luminet, J.P. Image of a spherical black hole with thin accretion disk. *Astron. Astrophys.* **1979**, *75*, 228.
8. Bardeen, J.M. Rapidly rotating stars, disks, and black hole. In *Black Holes (Les Astres Occlus)*; Dewitt, C., Dewitt, B.S., Eds.; Gordon and Breach Science Publishers: New York, NY, USA, 1973; pp. 215–239.
9. de Vries, A. The apparent shape of a rotating charged black hole, closed photon orbits and the bifurcation set A_4. *Class. Quantum Grav.* **2000**, *17*, 123. [CrossRef]
10. Hioki, K.; Maeda, K.I. Measurement of the Kerr spin parameter by observation of a compact object's shadow. *Phys. Rev. D* **2009**, *80*, 024042. [CrossRef]
11. Abdujabbarov, A.; Atamurotov, F.; Kucukakca, Y.; Ahmedov, B.; Camci, U. Shadow of Kerr-Taub-NUT black hole. *Astrophys. Space Sci.* **2012**, *344*, 429. [CrossRef]
12. Amarilla, L.; Eiroa, E.F.; Giribet, G. Null geodesics and shadow of a rotating black hole in extended Chern-Simons modified gravity. *Phys. Rev. D* **2010**, *81*, 124045. [CrossRef]
13. Amarilla, L.; Eiroa, E.F. Shadow of a rotating braneworld black hole. *Phys. Rev. D* **2012**, *85*, 064019. [CrossRef]
14. Amarilla, L.; Eiroa, E.F. Shadow of a Kaluza-Klein rotating dilaton black hole. *Phys. Rev. D* **2013**, *87*, 044057. [CrossRef]
15. Zhu, T.; Wu, Q.; Jamil, M.; Jusufi, K. Shadows and deflection angle of charged and slowly rotating black holes in Einstein-Æther theory. *Phys. Rev. D* **2019**, *100*, 044055. [CrossRef]
16. Haroon, S.; Jamil, M.; Jusufi, K.; Lin, K.; Mann, R.M. Shadow and deflection angle of rotating black holes in perfect fluid dark matter with a cosmological constant. *Phys. Rev. D* **2019**, *99*, 044015. [CrossRef]
17. Konoplya, R.A. Shadow of a black hole surrounded by dark matter. *Phys. Lett. B* **2019**, *795*, 1. [CrossRef]
18. Wei, S.-W.; Liu, Y.-X. Parametric study of Kerr black hole shadow: analytical and exact calculations. *J. Cosmol. Astropart. Phys.* **2013**, *11*, 063. [CrossRef]
19. Xu, Z.; Hou, X.; Wang, J. Possibility of identifying matter around rotating black hole with black hole shadow. *J. Cosmol. Astropart. Phys.* **2018**, *10*, 046. [CrossRef]
20. Hou, X.; Xu, Z.; Zhou, M.; Wang, J. Black hole shadow of Sgr A* in dark matter halo. *J. Cosmol. Astropart. Phys.* **2018**, *1807*, 015. [CrossRef]
21. Jusufi, K.; Jamil, M.; Salucci, P.; Zhu. T.; Haroon, S. Black Hole surrounded by a dark matter halo in the M87 galactic center and its identification with shadow images. *Phys. Rev. D* **2019**, *100*, 044012. [CrossRef]
22. Cunha, P.V.P.; Herdeiro, C.A.R.; Radu. E. EHT constraint on the ultralight scalar hair of the M87 supermassive black hole. *Universe* **2019**, *5*, 220. [CrossRef]
23. Amir, M.; Jusufi, K.; Banerjee, A.; Hansraj, S. Shadow images of Kerr-like wormholes. *Class. Quantum Grav.* **2019**, *36*, 215007. [CrossRef]
24. Amir, M.; Singh, B.P.; Ghosh, S.G. Shadows of rotating five-dimensional charged EMCS black holes. *Eur. Phys. J. C* **2018**, *78*, 399. [CrossRef]
25. Shaikh, R. Shadows of rotating wormholes. *Phys. Rev. D* **2018**, *98*, 024044. [CrossRef]
26. Shaikh, R.; Kocherlakota, P.; Narayan, R.; Joshi, P.S. Shadows of spherically symmetric black holes and naked singularities. *Mon. Not. R. Astron. Soc.* **2019**, *482*, 52. [CrossRef]
27. Gyulchev, G.; Nedkova, P.; Tinchev, V.; Yazadjiev, S. On the shadow of rotating traversable wormholes. *Eur. Phys. J. C* **2018**, *78*, 544. [CrossRef]
28. Gyulchev, G.; Nedkova, P.; Tinchev, V.; Yazadjiev, S. Cusp structure in shadows casted by rotating wormholes. *AIP Conf. Proc.* **2019**, *2075*, 040005.
29. Abdujabbarov, A.; Amir. M.; Ahmedov, B.; Ghosh, S.G. Shadow of rotating regular black holes. *Phys. Rev. D* **2016**, *93*, 104004. [CrossRef]
30. Abdujabbarov, A.; Juraev, B.; Ahmedov, B.; Stuchlík, Z. Shadow of rotating wormhole in plasma environment. *Astrophys. Space Sci.* **2016**, *361*, 226. [CrossRef]

31. Amir, M.; Banerjee, A.; Maharaj, S.D. Shadow of charged wormholes in Einstein-Maxwell-dilaton theory. *Ann. Phys.* **2019**, *400*, 198. [CrossRef]
32. Bambi, C.; Freese, K. Apparent shape of super-spinning black holes. *Phys. Rev. D* **2009**, *79*, 043002. [CrossRef]
33. Broderick, A.E.; Johannsen, T.; Loeb, A.; Psaltis, D. Testing the no-hair theorem with Event Horizon Telescope observations of Sagittarius A*. *Astrophys. J.* **2014**, *784*, 7. [CrossRef]
34. Johannsen, T.; Broderick, A.E.; Plewa, P.M.; Chatzopoulos, S.; Doeleman, S.S.; Eisenhauer, F.; Fish, V.L.; Genzel, R.; Gerhard, O.; Johnson, M.D. Testing general relativity with the shadow size of Sgr A*. *Phys. Rev. Lett.* **2016**, *116*, 031101. [CrossRef] [PubMed]
35. Mizuno, Y.; Younsi, Z.; Fromm, C.M.; Porth, O.; De Laurentis, M.; Olivares, H.; Falcke, H.; Kramer M.; Rezzolla, L. The current ability to test theories of gravity with black hole shadows. *Nat. Astron.* **2018**, *2*, 585. [CrossRef]
36. Younsi, Z.; Zhidenko, A.; Rezzolla, L.; Konoplya, R.; Mizuno, Y. New method for shadow calculations: Application to parametrized axisymmetric black holes. *Phys. Rev. D* **2016**, *94*, 084025. [CrossRef]
37. Wei, S.W.; Liu, Y.X.; Mann, R.B. Intrinsic curvature and topology of shadows in Kerr spacetime. *Phys. Rev. D* **2019**, *99*, 041303. [CrossRef]
38. Wei, S.W.; Zou, Y.C.; Liu, Y.X.; Mann, R.B. Curvature radius and Kerr black hole shadow. *J. Cosmol. Astropart. Phys.* **2019**, *1908*, 030. [CrossRef]
39. Kibble, T.W.B. Topology of cosmic domains and strings. *J. Phys. A* **1976**, *9*, 1387. [CrossRef]
40. Barriola, M.; Vilenkin, A. Gravitational field of a global monopole. *Phys. Rev. Lett.* **1989**, *63*, 341. [CrossRef]
41. Teixeira Filho, R.; Bezerra, B. Gravitational field of a rotating global monopole. *Phys. Rev. D* **2001**, *64*, 084009. [CrossRef]
42. Jusufi, K.; Werner, M.C.; Banerjee, A.A.; Ovgun, A. Light deflection by a rotating global monopole spacetime. *Phys. Rev. D* **2017**, *95*, 104012. [CrossRef]
43. Ono, T.; Ishihara, A.; Asada, H. Deflection angle of light for an observer and source at finite distance from a rotating global monopole. *Phys. Rev. D* **2019**, *99*, 124030. [CrossRef]
44. Newman, E.T.; Janis, A.I. Note on the Kerr Spinning-Particle Metric. *J. Math. Phys.* **1965**, *6*, 915. [CrossRef]
45. Dutta, S.; Jain, A.; Soni, R. Dyonic black hole and holography. *J. High Energy Phys.* **2013**, *2013*, 60. [CrossRef]
46. Erbin, H. Janis–Newman Algorithm: Generating rotating and NUT charged black holes. *Universe* **2017**, *3*, 19. [CrossRef]
47. Heydarzade, Y.; Darabi, F. Black hole solutions surrounded by perfect fluid in Rastall theory. *Phys. Lett. B* **2017**, *771*, 365–373. [CrossRef]
48. Azreg-Aïnou, M. Generating rotating regular black hole solutions without complexification. *Phys. Rev. D* **2014**, *90*, 064041. [CrossRef]
49. Haroon, S.; Jamil, M.; Lin, K.; Pavlovic, P.; Sossich, M.; Wang, A. The effects of running gravitational coupling on rotating black holes. *Eur. Phys. J. C* **2018**, *78*, 519. [CrossRef]
50. Dadhich, N.; Narayan, K.; Yajnik, U.A. Schwarzschild black hole with global monopole charge. *Pramana* **1998**, *50*, 307. [CrossRef]
51. Plebanski, J.F.; Demianski, M. Rotating, charged, and uniformly accelerating mass in general relativity. *Ann. Phys.* **1976**, *98*, 98. [CrossRef]
52. Grenzebach, A.; Perlick, V.; Lämmerzahl, C. Photon regions and shadows of Kerr-Newman-NUT black holes with a cosmological constant. *Phys. Rev. D* **2014**, *89*, 124004. [CrossRef]
53. Chandrasekhar, S. *The Mathematical Theory of Black Holes*; Oxford University Press: New York, NY, USA, 1992.

© 2020 by the authors. Licensee MDPI, Basel, Switzerland. This article is an open access article distributed under the terms and conditions of the Creative Commons Attribution (CC BY) license (http://creativecommons.org/licenses/by/4.0/).

Article

EHT Constraint on the Ultralight Scalar Hair of the M87 Supermassive Black Hole

Pedro V. P. Cunha [1,2,3,*,†], Carlos A. R. Herdeiro [1,2,3,†] and Eugen Radu [1,2,†]

1. Departamento de Matemática da, Universidade de Aveiro, Campus de Santiago, 3810-193 Aveiro, Portugal; herdeiro@ua.pt (C.A.R.H.); eugen.radu@ua.pt (E.R.)
2. Centre for Research and Development in Mathematics and Applications (CIDMA), Campus de Santiago, 3810-183 Aveiro, Portugal
3. Centro de Astrofísica e Gravitação—CENTRA, Departamento de Física, Instituto Superior Técnico—IST, Universidade de Lisboa—UL, Av. Rovisco Pais 1, 1049-001 Lisboa, Portugal
* Correspondence: pintodacunha@tecnico.ulisboa.pt
† These authors contributed equally to this work.

Received: 20 September 2019; Accepted: 23 November 2019; Published: 27 November 2019

Abstract: Hypothetical ultralight bosonic fields will spontaneously form macroscopic bosonic halos around Kerr black holes, via superradiance, transferring part of the mass and angular momentum of the black hole into the halo. Such a process, however, is only efficient if resonant—when the Compton wavelength of the field approximately matches the gravitational scale of the black hole. For a complex-valued field, the process can form a stationary, bosonic field black hole equilibrium state—a black hole with synchronised hair. For sufficiently massive black holes, such as the one at the centre of the M87 supergiant elliptic galaxy, the hairy black hole can be robust against its own superradiant instabilities, within a Hubble time. Studying the shadows of such scalar hairy black holes, we constrain the amount of hair which is compatible with the Event Horizon Telescope (EHT) observations of the M87 supermassive black hole, assuming the hair is a condensate of ultralight scalar particles of mass $\mu \sim 10^{-20}$ eV, as to be dynamically viable. We show the EHT observations set a weak constraint, in the sense that typical hairy black holes that could develop their hair dynamically, are compatible with the observations, when taking into account the EHT error bars and the black hole mass/distance uncertainty.

Keywords: ultralight particles; black hole shadow; event horizon telescope

---

1. Introduction

The hypothesis that all astrophysical black holes (BHs) when near equilibrium are well described by the Kerr metric [1]—*the Kerr hypothesis*—yields a remarkable scenario. It means that throughout the whole mass spectrum, ranging from solar mass BHs, with $M \sim M_\odot$, all the way until the most supermassive black holes known, with $M \sim 10^{10} M_\odot$, the immense population of astrophysical BHs correspond to the very same object, with only two macroscopic degrees of freedom. One of these is the mass, which merely rescales the BH, leaving a single degree of freedom with impact on the BH phenomenology—the spin. The Kerr hypothesis, therefore, encodes an economical natural order: The landscape of gravitational atoms (BHs) that compose the dark Universe is made up of a single species, varying only in size (by, at least, ten orders of magnitude!) and spin, but otherwise indistinguishable. Such uniformity is a trademark of the microscopic world, where all elementary particles of a given species are indistinguishable, but not of the macroscopic world, where variety is ubiquitous.

Despite the current lack of tension between observations and the Kerr hypothesis, there are reasons to consider the latter but a fair approximation, within current precision, rather than a fundamental truth. Both fundamental problems—such as a quantum version of the laws of gravity, and how it impacts on the physics of horizons and classical singularities—and the phenomenological problem of accounting for the elusive dark matter and dark energy, suggest our current understanding of gravity is incomplete. It may therefore be that the Kerr hypothesis is strictly false for astrophysical BHs at all scales. But an alternative possibility is that the Kerr hypothesis is violated at a higher degree at some narrow interval of scales only, wherein new physics exists, remaining an excellent approximation outside this interval.

A concrete realisation of the latter possibility is provided by scenarios of hypothetical ultralight bosonic particles that could constitute part of the dark matter population [2–4]. Inspired by the QCD axion [5], and with theoretical support in string theory [6], these scenarios suggest that a landscape of such particles, negligibly interacting with standard model constituents, might exist, leaving their gravitational interactions as the only possible smoking gun for their identification. Amongst these, an exciting possibility, is their interaction with BHs, which, could single out a scale (or a range of scales) wherein BHs deviate from the Kerr paradigm.

Spinning BHs (in particular Kerr BHs) are energy and angular momentum reservoirs that can be classically mined. A very well suited tool for such mining is an ultralight bosonic field. Then, through the phenomenon of superradiance [7,8] an appropriate small seed of such field (provided, say, as a quantum fluctuation) will grow into a macroscopic condensate of bosonic particles—a Bose–Einstein condensate—storing a non-negligible fraction of the original BH mass and angular momentum. When the energy/angular momentum transfer from the BH to the bosonic halo saturates, the corresponding BH-halo state may or may not be stationary. If the bosonic field is real, the BH-halo system emits gravitational radiation and slowly decays back to a Kerr BH [9]. But if the bosonic field is complex, the BH-halo system is stationary, which is, in fact, a hairy BH—dubbed BHs with synchronised hair [10].

BHs with synchronised hair are not absolutely stable. They are themselves prone to their own superradiant instabilities [11–13]. However, the timescale of these instabilities is larger than the one of the initial Kerr superradiant instability that formed the hair, and, in the right mass range it becomes cosmologically large. A suggestive possibility is then the following. A Kerr BH develops ultralight bosonic hair in an astrophysical time scale; the hairy BH is then effectively stable, since it is superradiantly unstable only in a cosmological timescale. This turns out to be a realisable scenario for supermassive BHs, such as the one recently observed by the Event Horizon Telescope (EHT) collaboration [14–16] at the centre of the supergiant elliptic galaxy M87, henceforth referred to as the M87 BH.

The scenario in this paper is therefore that the M87 BH is hairy, due to an appropriate ultralight scalar field. Appropriate means its mass is in the right range to make the superradiant instability of the original Kerr BH grow in a sufficiently small fraction of the Hubble time, yielding a hairy BH that is stable in a cosmological time scale. Since the shadows of a Kerr and a hairy BH with the same total mass and angular momentum differ [17], and since the EHT observation is compatible with the M87 BH being of Kerr type, we shall then inquire how much the EHT observations constrain the hair. As we shall see, for the most interesting mass ranges, as to make the hairy BH dynamically viable, the EHT constraint is weak, and it is essentially compatible with a hairy BH that could have dynamically formed from superradiance and it is in a long lived, albeit not absolutely stable, state.

This paper is organised as follows. In Section 2 we discuss the physical scenario under which a BH with synchronised scalar hair could form from superradiance and be effectively stable within a cosmological timescale. In Section 3 we describe the part of the domain of existence of hairy BHs that is dynamically viable, according to the criteria in Section 2 and that we shall study in the remaining of the paper. In Section 4 we consider the Kerr BH shadow and we obtain an approximate expression for the shadow areal radius, valid for arbitrary observation angle and dimensionless spin value. In Section 5 we analyse the shadows of the hairy BHs in the relevant domain and obtain an approximate expression for the areal shadow radius, in terms of that of a comparable Kerr BH, i.e., with the same mass, and a parameter measuring the hairiness of the BH. In Section 6 the expression obtained in Section 5 is applied to the case of the M87 supermassive BH. Then, considering the EHT observational errors, together with the errors in the mass estimate, we constrain the hairiness compatible with the observations. Brief final remarks are given in Section 7. Unless otherwise stated, natural units $c = 1 = G = \hbar$ are used.

2. The Hair Formation and Hair Instability Timescales

The timescale of BH superradiance is extremely sensitive to the occurrence, or not, of a resonance between the gravitational scale of the BH and the Compton wavelength of the ultralight particles. Consider a massive, complex, scalar field, Φ, described by the Klein–Gordon equation, $\Box \Phi = \mu^2 \Phi$, with mass μ, on the background of a Kerr BH with mass M. Maximal efficiency occurs for $M\mu \simeq 0.4$ [18] and for a spin close to extremality; for the M87 BH, for which we take $M_{M87} \sim 6 \times 10^9 M_\odot$[1] this resonant scalar field mass is:

$$\mu_r = \frac{0.4}{M_{M87}} \simeq 4 \times 10^{-20} \text{ eV} . \tag{1}$$

At maximal efficiency, the e-folding time of the superradiant instability's leading mode is [18]:

$$\Delta t \sim 10^7 \tau_{LC} , \tag{2}$$

where τ_{LC} is the light crossing time of the BH. For the M87 BH, $\Delta t \sim 10^4$ years. We call this the *hair formation timescale*. This means that if an ultralight boson of mass μ_r exists, the M87 BH would develop scalar hair in an astrophysically short time scale.

If the resonance $\mu = \mu_r$ is missed, however, this timescale grows extremely fast: as $(M\mu)^{-8}$ [23] for the leading mode and $M\mu \ll 1$; or as $10^7 e^{3.7 M\mu}$ for $M\mu \gg 1$ [9,24]. In other words, if $M\mu$ misses the resonant sweet spot by one order of magnitude, either above or below, the timescale of the leading mode of the superradiant instability of Kerr becomes considerably larger than the Hubble time, and the Kerr BH does not become hairy. On the lower end, the hair formation timescale becomes larger than one tenth of the Hubble time for $M\mu < 0.05$ [13], for the M87 BH mass. Thus, a conservative bound on the formation timescale is to take $M\mu > 0.1$, when the formation timescale becomes around a thousandth of the Hubble time.

Supermassive BHs in matter rich environments, such as galactic centres, are expected to grow in time. Thus, a supermassive BH with $\sim 10^9 M_\odot$, such as the one at the centre of M87, will have evolved, due to accretion and mergers, from one (or many) BHs with mass several orders of magnitude lower, see e.g., [25]. Only when the BH grows to the size that resonates with $\sim \mu_r$ does the superradiant energy/angular momentum extraction becomes efficient producing a sufficiently non-Kerr BH. At all other scales BHs remain Kerr-like.

Once the hairy BH forms, one must consider its leading superradiant instability. The leading instability has an e-folding time, dubbed *hair instability timescale*, larger than the Hubble time if $M\mu \lesssim 0.25$ [13]. Thus, a Kerr BH with mass $M = M_{M87}$ becomes hairy in an astrophysical timescale and the hair is stable in a cosmological timescale if (conservatively rounding off 0.25 to 0.3):

$$\mu M_{M87} \in [0.1, 0.3] \quad \Rightarrow \quad \mu \in [1, 3] \times 10^{-20} \text{ eV} . \tag{3}$$

We remark that there is a dependence on the BH dimensionless spin parameter a in determining the resonant scalar field mass μ_r, although the instability is considerably more sensitive to $M\mu$ than a. If the spin is not near extremal, this changes the ideal value of $M\mu$ given in Equation (1), which could

---

[1] For the considerations in this section this approximate value suffices. More accurate values will be considered in Section 6. This value is suggested from stellar dynamics [19] and favoured by the EHT observations [14]. A value half of this is suggested by gas dynamics [20]. The spin of the M87 BH is largely unknown, with different claims in the literature, see e.g., [21,22].

push down slightly, but not significantly, the lower end value of the interesting mass range given in Equation (3) (see also discussion in Section 7).

How much energy can be extracted from the Kerr BH region into the hair? Fully non-linear numerical evolutions of the superradiant instability of a Kerr BH triggered by a complex *vector* field were performed in [26], leading to the formation of that BH with synchronised (vector) hair [27], first discussed in [28]. These simulations showed that the maximal energy extracted was $\sim 0.09M$ of the original BH with mass M. The largest energy extraction, moreover, occurred for the lowest values of $M\mu$ ($M\mu = 0.25$, for the simulations in [26]), for which the superradiant growth was slower. The trend moreover suggests that the 9% may be close to the maximal possible value, in the vector case. The corresponding value in the scalar case is unknown. Since the process is slower in the scalar case, it is conceivable that larger energy extractions are possible—see [29]. But, in any case, thermodynamic sets a limit of 29% to the rotational energy that can be extracted from a Kerr BH.

3. The Selected Part of the Domain of Existence

The full domain of existence of BHs with synchronised scalar hair was obtained in [10,30]. These are stationary and axisymmetric solutions of the Einstein–(massive, complex)Klein–Gordon system. The shadows of these BHs have been explored in [17,31,32]. However, a detailed study of the shadow properties in the region of dynamical interest unveiled in the previous section remains to be done. The goal of the reminder of this paper is to perform such study and relating it to the EHT observations.

The part of the domain of existence describing the astrophysically viable solutions, in relation to the M87 BH, as described in the last section, corresponds to values of $M\mu$ in accordance with Equation (3). For $M\mu \lesssim 0.1$, obtaining numerically the hairy BH solutions becomes challenging, due to the different scales involved in the problem. So, we shall perform our analysis of the shadows in a section of the domain of existence for $0.2 \leqslant M\mu \leqslant 0.5$, which allowed us to obtain interpolations with more data. The main conclusions are not substantially affected by this choice of sample. As we shall see, the obtained trend is already informative.

In Figure 1 we exhibit the configurations analysed to obtain the behaviour of the shadows of the hairy BHs in the dynamically viable region.

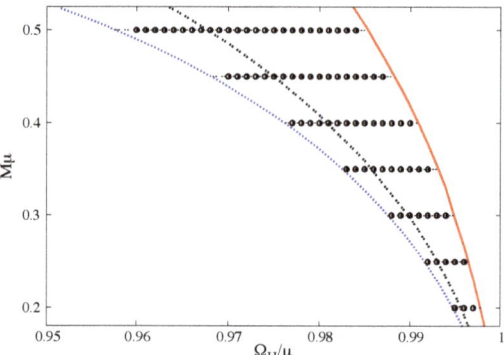

Figure 1. The section of the domain of existence of hairy black holes (BHs) to be analysed. We have chosen sequences of representative solutions with constant $M\mu$—black dots. Their shadow properties are analysed and the corresponding trends interpolated for the whole region. The dashed-dotted line separates solutions with more (to the right) and less (to the left) than 29% of the spacetime energy in the scalar hair.

They are displayed in the configuration space (Ω_H/μ, $M\mu$), with Ω_H and μ being, respectively, the BH horizon angular velocity and the boson mass. The synchronisation condition[2] means that $\omega = \Omega_H$ for this family of solutions, where ω is the frequency of the complex scalar field, that oscillates harmonically (but with a time independent energy momentum tensor). The Kerr limit, in which the hair vanishes, is provided by the dotted blue line; the rotating boson star limit, in which the horizon vanishes, is provided by the solid red line.

As an alternative to the $\{\Omega_H/\mu, M\mu\}$ labelling of this section of the domain of existence of the hairy BHs, each solution can also be labelled by the pair $\{p, M\mu\}$, where [33]:

$$p \equiv 1 - \frac{M_H}{M}, \qquad (4)$$

M_H is the horizon energy (measured by a Komar integral) and M the ADM energy. Thus, p measures the fraction of the spacetime energy in the hair. This parameter satisfies $0 \leqslant p \leqslant 1$; $p = 0$ ($p = 1$) corresponds to the Kerr (boson star) limit, displayed as the dotted blue (solid red) lines in Figure 1. In the figure, a dotted-dashed black line denotes the $p = 0.29$ threshold, above which the hairy BHs cannot form from the superradiant instability of Kerr. Nevertheless, configurations with $p > 0.29$ are also equally valid equilibrium solutions to the Einstein–Klein–Gordon field equations. Although at the moment this is unclear, these configurations could perhaps form by alternative channels other than the superradiant instability. For the analysis being performed, considering configurations with $p > 0.29$ increases the sample size for the shadow interpolations in Section 5.

4. The Kerr BH Shadow

The calculation of the BH shadow for a Schwarzschild and Kerr BH was introduced by Synge [34] and Bardeen [35] respectively. A pioneering computation in an astrophysical environment was done by Luminet [36,37] and its measurability was first assessed in [38]—see [39] for a review and [40] for a working setup to compute shadows via ray tracing.

Given an observational setup, the observer can measure the BH shadow image[3] area $\mathcal{A}$. We define the shadow *areal radius* as:

$$S \equiv \sqrt{\frac{\mathcal{A}}{\pi}}, \qquad (5)$$

which is well defined even for non-circular shadow shapes. In what follows, the shadow radius[4] S will be compared between hairy and Kerr BHs. The latter is known analytically in particular cases, as we shall now review for our subsequent application.

4.1. Two Cases for Which the Kerr Shadow Areal Radius Is Exactly Computable

For an observer at infinity, the Kerr shadow edge is known analytically for all spin values $a = J/M$ [39], where J is the total angular momentum of the spacetime and the existence of a horizon requires $0 \leqslant a^2 \leqslant M^2$. From this analytic knowledge of the shadow edge, however, it might not always be possible to compute $\mathcal{A}$ exactly, and hence the Kerr shadow areal radius, denoted $S_{\text{Kerr}}(a, \theta_o)$, which generically depends on a and the polar angle of the observer θ_o. But in two special cases this is possible.

The first case is when the observer is on the rotation axis, i.e., when $\theta_o = \{0, \pi\}$. Then, the Kerr shadow edge is circular due to axial-symmetry. In this case the shadow radius $S_{\text{Kerr}}(a, \text{axis})$ is fully determined by a zero angular momentum spherical photon orbit with a Boyer–Linquist radial

---

[2] The synchronization condition is an equilibrium requirement on the existence of these hairy BH solutions (i.e., within the valid domain in Figure 1), and it is not directly used in the rest of the analysis.
[3] The shadow in the image domain is rescaled with respect to its angular size ϑ by a factor $\mathcal{R}$, see Equation (12).
[4] Other possible measures for the shadow size can be found in the literature, e.g., see [41].

coordinate r. For this special case $\mathcal{A}$ can be obtained exactly and so can the shadow areal radius. Using the results in [42], the latter is obtained to be, as a function of a:

$$S_{\text{Kerr}}(a, \text{axis}) = \sqrt{\chi + a^2}, \qquad (6)$$

where:

$$\chi = r^2 \left(\frac{3r^2 + a^2}{r^2 - a^2} \right), \quad \frac{r}{M} = 1 + 2\sqrt{1 - \frac{a^2}{3}} \cos\left[\frac{1}{3} \arccos\left(\frac{1 - a^2}{\sqrt{\left(1 - \frac{a^2}{3}\right)^3}} \right) \right].$$

Observe that for $a = 0$, then $r = 3M$, $\chi = 27M^2$ and $S_{\text{Kerr}}(0, \text{axis}) = \sqrt{27}M$, which are the familiar Schwarzschild light ring coordinate, the corresponding impact parameter (squared) and shadow areal radius.

The second case is for an extremal Kerr BH (i.e., $|a| = M$), viewed from the equatorial plane, i.e., with $\theta_o = \pi/2$. Then, the shadow edge is not circular. However, the shadow shape $y(x)$ in Cartesian-like coordinates (x, y) simplifies into [39] (in units of M):

$$y(x) = \pm\sqrt{11 + 2x - x^2 + 8\sqrt{2 + x}}, \qquad x \in [-2, 7],$$

in which case, the area $\mathcal{A}$ can be explicitly computed:

$$\mathcal{A} = \int_{-2}^{7} 2y(x)\, dx = (15\sqrt{3} + 16\pi) M^2,$$

which leads to a shadow radius:

$$S_{\text{Kerr}}\left(\pm M, \frac{\pi}{2}\right) = \sqrt{16 + \frac{15\sqrt{3}}{\pi}} M. \qquad (7)$$

4.2. An Approximation for the Areal Radius of the Kerr Shadow

We were not able to find an exact expression for the Kerr shadow areal radius in the generic case. We have verified, however, that as seen by an observer at infinity, $S_{\text{Kerr}}(a, \theta_o)$ can be estimated (within an error $\lesssim 0.8\%$) as:[5]

$$S_{\text{Kerr}}(a, \theta_o) \simeq S_{\text{Kerr}}(a, \text{axis}) + \frac{2|a|\theta_o}{\pi M} \left[S_{\text{Kerr}}\left(M, \frac{\pi}{2}\right) - S_{\text{Kerr}}(M, \text{axis}) \right], \qquad (8)$$

where $S_{\text{Kerr}}(M, \text{axis}) = (2 + 2\sqrt{2})M$. This approximation will be used in the following.

To make contact with the Kerr limit in the domain of existence displayed in Figure 1, we observe that from the points along the blue dotted line therein, the Kerr spin $|a|$ can be obtained from $M\mu$ using:

$$a = \frac{M^2 \Omega_H}{M^2 \Omega_H^2 + 1/4}, \qquad M\Omega_H \simeq b_1 + b_2 M\mu + b_3 M^2 \mu^2,$$

where the first expression is exact for Kerr BHs, and the second is a good approximate relation along the Kerr existence (blue) line in Figure 1, with parameters:

$$(b_1, b_2, b_3) = (-0.00926172, 1.08238, -0.209874). \qquad (9)$$

---

[5] This error was determined through a direct comparison of the approximate formula with the corresponding Kerr values. Recall that the Kerr shadow edge is known analytically and determining the shadow areal radius amounts to solving, numerically, the area integral. Thus the Kerr shadow areal radius, albeit obtained numerically, is computed with a precision considerably better than 0.8%.

Thus, for the Kerr BHs in Figure 1, the shadow areal radius becomes a function of $M\mu$ and θ_o, $S_{\text{Kerr}}(a(M\mu), \theta_o)$.

5. Hairy BHs Shadow in the Considered Domain of Existence

Since the shadows of the hairy BHs were obtained through a numerical ray tracing procedure [40], they correspond to an observer at finite, rather than infinite, distance from the BH. The observer is placed at a finite perimetral distance $\mathcal{R} = \sqrt{g_{\varphi\varphi}}(r_o, \pi/2)$, where $\partial/\partial\varphi$ is the Killing vector field associated to axial-symmetry. The quantity $\mathcal{R}$ is defined for each radial coordinate r_o. For two observers $\{\mathcal{O}_1, \mathcal{O}_2\}$, respectively with $\{\mathcal{R}_1, \mathcal{R}_2\} \gg M$ and corresponding shadow radii $\{S_1, S_2\}$, a simple extrapolation can provide a good approximation for the shadow radius S_∞ at infinity:

$$S_\infty \simeq S_2 - \left(\frac{S_2 - S_1}{1 - \mathcal{R}_2/\mathcal{R}_1}\right).$$

In our setup, we take $\mathcal{R}_1 = 100M$ and $\mathcal{R}_2 = 200M$.

Hairy BH Shadow Approximation for $\theta_o = 17°$

The angle between the M87 BH spin and the line of sight has been estimated to be 17° from the observed jet [43]. Choosing, then, $\theta_o = 17°$ in order to compare with EHT's M87 observation, the shadow's areal radius at infinity of the hairy BHs, $S_{\text{hairy}}(p, M\mu, \theta_o)$ can be approximated within an error $\lesssim 0.8\%$ as:

$$S_{\text{hairy}}(p, M\mu, 17°) \simeq (1-p)\left(S_{\text{Kerr}}(a(M\mu), 17°) + \beta_1 p \, M^2\mu\right), \qquad \beta_1 \simeq 1.21455, \qquad (10)$$

The accuracy of this approximation is clear from Figure 2, where both the shadow radius of the hairy BHs and the function $S_{\text{hairy}}(p, M\mu, 17°)$, are exhibited.

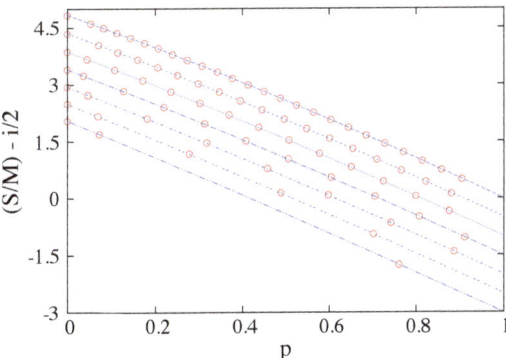

Figure 2. Areal shadow radius for hairy BHs and the analytic approximation of Equation (10). Circles correspond to data for the individual solutions in Figure 1. Each straight line is a set with constant $M\mu$ in Figure 1, given by $M\mu = 0.5 - i/20$ and $i = \{0, \cdots, 6\}$ as we go from the top to the bottom straight line. The function exhibited in the y-axis is $S_{\text{hairy}}(p, M\mu, 17°)/M - i/2$, as to more clearly distinguish the different lines, where i is a function of $M\mu$ via $i = 10 \times (1 - 2M\mu)$.

The analysis of the hairy BHs data, leading to Equation (10), shows that the shadow's areal radius relative deviation δS, between hairy and Kerr BHs, depends very weakly on $M\mu$. It is therefore accurately parameterized by a function of p only:

$$\delta S(p) \equiv 1 - \frac{S_{\text{hairy}}(p, M\mu, 17^\circ)}{S_{\text{Kerr}}(a(M\mu), 17^\circ)} \simeq p + p(p-1)A, \qquad \text{with} \quad A \simeq 0.111159. \qquad (11)$$

This approximation is represented in Figure 3 as a solid line, together with the corresponding data (red circles), showing a very good agreement.

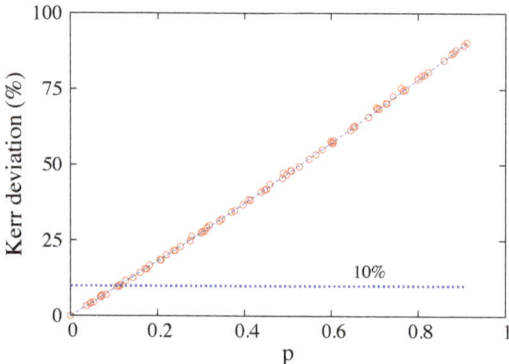

Figure 3. Deviation δS between the hairy BHs shadow areal radius and that of Kerr BHs, as function of p, for $\theta_o = 17^\circ$. The fit function $p + A\,p(p-1)$ with $A \simeq 0.111159$ captures the main features of δS.

A rough conclusion from this analysis is that a Kerr deviation no larger than $\sim 10\%$ is compatible with a $p \lesssim 11\%$ (dotted line in Figure 3). Indeed, the EHT measurement of the M87 BH shadow has an error bar of around 10% as discussed in the next section where a more precise statistical analysis is performed. To make contact with the observations, we note that for a dimensionless areal radius S/M, the corresponding angular radius in the sky is:

$$\vartheta = (S/M)\frac{M}{\mathcal{R}}. \qquad (12)$$

This relates theory with observation. Using it we will now restrict the values of p of a hairy BH that may be compatible with the EHT M87 BH observation data.

6. Application to the M87 BH Shadow

The EHT observation measures the emission ring *diameter* to be 42 ± 3 μas [14]. As discussed in [16], this emission ring diameter is not simply assumed to be the edge of the shadow. Calibration between the emission ring diameter and the photon ring (determining the edge of the shadow) based on GRMHD simulations leads to a 10% offset between the two. Although this offset is estimated from Kerr, it is conceivable that a similar effect arises for hairy BHs in the region of interest of Figure 1, which are still very much Kerr-like. Thus, we assume the M87 BH shadow diameter is 10% smaller than the EHT's observed emission ring, leading to an observed angular size of the shadow (corresponding to the areal *radius*) of:

$$\vartheta_o = (18.9 \pm 1.5)\,\mu\text{as}. \qquad (13)$$

The error bars in Equation (13) already provide some margin to accommodate a non-Kerr BH with the same mass. But a proper analysis must, in addition, take into account the error in the mass measurement, which must be an independent measurement from the EHT observations. As discussed in [16], both the independent measurements of the M87 BH mass, by Gebhardt et al. [19] based on star dynamics, and Walsh et al. [20] based on gas motion, actually directly measure the ratio $\lambda = M/L$, rather than M, where L is the luminosity distance, that we identify with $\mathcal{R}$. The mass in these works is then obtained *assuming* $L = 17.9$ Mpc, since the relative error for the distance is smaller. Taking their reported values for M and inferring the associated ratios for that distance, one has:

Gebhardt et al. (star motion): $\quad M = (6.6 \pm 0.4) \times 10^9 M_\odot , \quad \lambda = 0.369 \pm 0.022 \left(\dfrac{10^9 M_\odot}{\text{Mpc}} \right) ,$

Walsh et al. (gas motion): $\quad M = 3.5^{+0.9}_{-0.7} \times 10^9 M_\odot , \quad \lambda = 0.196^{+0.05}_{-0.04} \left(\dfrac{10^9 M_\odot}{\text{Mpc}} \right) .$

Choosing either of these data sets, we can now analyse the domain in the (p, λ) plane that provides an angular shadow size; using Equations (10) and (12) yields:

$$\vartheta = \lambda \frac{S_{\text{hairy}}(p, M\mu, 17^\circ)}{M} \simeq \lambda (1-p) \left(\frac{S_{\text{Kerr}}(a(M\mu), 17^\circ)}{M} + \beta_1 p\, M\mu \right), \qquad (14)$$

consistent with the EHT shadow, within a certain number of standard deviations. This analysis is performed in Figure 4.

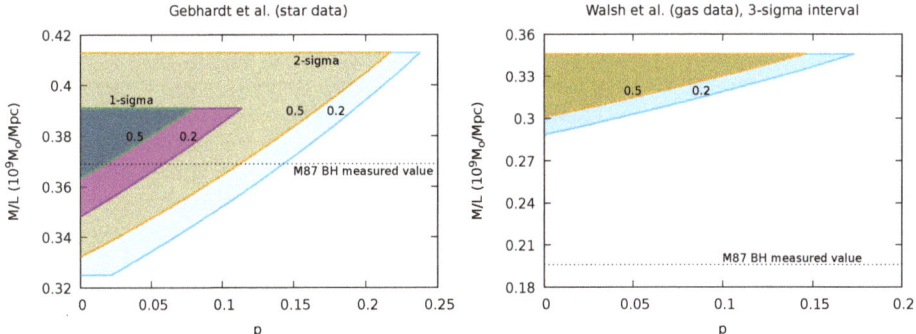

Figure 4. $(p, \lambda = M/L)$ domain providing values of $\vartheta = \lambda S_{\text{hairy}}(p, M\mu, 17^\circ)/M$ consistent with the Event Horizon Telescope (EHT) observation. (Left panel) within one and two standard deviations for the star motion data. (Right panel): within three standard deviations for the gas data. Small numbers indicate value of $M\mu$.

The left panel considers the star motion data. The shaded regions encode values of (p, λ) consistent with the EHT shadow observation, within one and two standard deviations, for the limiting values of $M\mu$ in the sample of solutions analysed. We conclude that within one standard deviation, $0 < p < 0.12$, and within two standard deviations $0 < p < 0.24$, for $M\mu = 0.2$. Slightly more restrictive values hold for $M\mu = 0.5$. The right panel considers the gas data. In this case, values in the (p, λ) domain exhibited can only agree with the EHT observation within 3 standard deviations of the observation error.

Considering the star dynamics data within one standard deviation, a hairy BH with $p \lesssim 0.12$ is compatible with the EHT observations. The trend with $M\mu$ in Figure 4 indicates, moreover, that for even lower values of $M\mu$—recall Equation (3)—the accommodated p is even slightly larger. Taking the simulations with a vector field discussed in Section 2 as an estimate of how much energy could be extracted dynamically into the hair, the tentative conclusion (with the caveat that the precise maximum amount of energy extractable dynamically is unknown in the scalar case) is that dynamically viable hairy BHs are compatible with the EHT observations, given the error bars.

The gas data, on the other hand, disfavours the hairy BHs, which is manifest in the mostly empty right panel of Figure 4, but it is also at some tension with the Kerr model, from the EHT observations. Within two standard deviations the data is incompatible with the model.

7. Final Remarks

With the advent of the first observation of a BH shadow by the EHT collaboration [14–16], a new direct window has now been opened into the strong gravity regime surrounding BHs. Together with the recent breakthroughs in gravitational wave astrophysics [44,45], and the precision upgrades that are expected to follow, the shadow observation opens the tantalizing possibility of testing existing BH models with direct observations.

In this paper, we have considered the possibility that the M87 supermassive BH has ultralight synchronised hair. We have made the case that some of these hairy BHs could be dynamically viable as a model of such a supermassive BH. Moreover, we have shown that the current EHT data, when taken together with the most favoured independent measurement of the mass of the M87 BH is compatible with the estimated range for dynamically viable hair. Thus such a hairy BH could be mistaken by a Kerr BH within all current measurements. See [43,46–50] for other constraints on ultralight dark matter from EHT data and, e.g., [51–54] for the impact of these data on other scenarios for M87.

Our analysis contains assumptions and some possible caveats, including:

- In Equation (1) we considered the resonant mass corresponding to the most efficient superradiant scenario, which in particular assumes a near extremal Kerr BH. If the spin is not near extremal (i.e., ideal to make superradiance as efficient as possible) this changes the ideal value of $M\mu$ given in Equation (1) in the text and, most importantly, it reduces the efficiency of the process and increases the timescale—see Figure 6 of [18]. As the dimensionless spin of the Kerr BH varies from 0.5 to 0.999, the timescale at maximal efficiency can vary by almost four orders of magnitude. This still allows the formation of scalar hair in less than 1% of a Hubble time in the M87 case: for maximal efficiency the time scale was 10^4 years for the M87 mass. This variation in the most efficient $M\mu$ could push down slightly, but not significantly, the lower end value of the interesting mass range given in Equation (3).
- Although we have considered that the most interesting mass interval in the context of our analysis is given by Equation (3), the analysis of hairy BH solutions was performed in a different mass range, cf. Figure 1. This was justified in Section 3 and we believe the main conclusions are not substantially affected by this choice of sample.
- Our work assumes the scalar hair around M87 is truly stationary, described by a minimally coupled massive, complex scalar field and forms from superradiance. If other mechanisms can form hairier BHs, or for other sorts of BHs with scalar hair (even if only approximately stationary), our conclusions do not apply, as, for example, in the scenario discussed in [47,55].
- In this paper we have used a single number (the shadow aerial radius) to set constraints. Other shadow measures could also be introduced (e.g., shadow deviation from a circle). However, due to the precision of the EHT measurement, such quantities would be too poorly constrained, at the moment. Such an analysis will be certainly interesting when more precise observations become possible.
- We have assumed that the M87 BH spin makes an angle of 17° with the line of sight, as suggested from the jet [43] and also assumed by the EHT analysis.
- We have assumed that there is an offset of about 10% between the size of the photon ring and the emission ring observed by the EHT. For Kerr this is justified by numerical GRMHD simulations—see also [56,57] for a discussion on this point. Since the hairy BHs in the region of interest are not very hairy, it is conceivable this offset is of a similar order.

- The gas data [20] was included in our discussion for completeness, as it was in the EHT paper VI [16]. However, this data is under tension even with the Kerr hypothesis, as discussed in detail in [16]. If the gas observations were to hold, they would have major implications concerning the Kerr paradigm. The conclusion that could be extracted here from this data is not different from the EHT paper: it is in tension with the models that were considered (including Kerr).

It would be very interesting to repeat the current analysis for the case of BHs with ultralight synchronised vector hair or scalar hair with self-interactions.

Author Contributions: All three authors contributed equally to the conceptualization, methodology, software, validation, formal analysis, investigation, resources, data curation, writing–original draft preparation, writing–review and editing, visualization, supervision, project administration, funding acquisition.

Funding: P.V.P.C. is supported by Grant No. PD/BD/114071/2015 under the FCT-IDPASC Portugal Ph.D. program. This work is supported by the Fundação para a Ciência e a Tecnologia (FCT) project UID/MAT/04106/2019 (CIDMA), by CENTRA (FCT) strategic project UID/FIS/00099/2013, by national funds (OE), through FCT, I.P., in the scope of the framework contract foreseen in the numbers 4, 5 and 6 of the article 23, of the Decree-Law 57/2016, of 29 August, changed by Law 57/2017, of 19 July. We acknowledge support from the project PTDC/FIS-OUT/28407/2017. This work has further been supported by the European Union's Horizon 2020 research and innovation (RISE) programmes H2020-MSCA-RISE-2015 Grant No. StronGrHEP-690904 and H2020-MSCA-RISE-2017 Grant No. FunFiCO-777740. The authors would like to acknowledge networking support by the COST Action CA16104.

Conflicts of Interest: The authors declare no conflict of interest.

References

1. Kerr, R.P. Gravitational field of a spinning mass as an example of algebraically special metrics. *Phys. Rev. Lett.* **1963**, *11*, 237–238. [CrossRef]
2. Suárez, A.; Robles, V.H.; Matos, T. A Review on the Scalar Field/Bose-Einstein Condensate Dark Matter Model. *Astrophys. Space Sci. Proc.* **2014**, *38*, 107–142.
3. Hui, L.; Ostriker, J.P.; Tremaine, S.; Witten, E. Ultralight scalars as cosmological dark matter. *Phys. Rev.* **2017**, *D95*, 043541. [CrossRef]
4. Visinelli, L.; Vagnozzi, S. Cosmological window onto the string axiverse and the supersymmetry breaking scale. *Phys. Rev.* **2019**, *D99*, 063517. [CrossRef]
5. Peccei, R.D.; Quinn, H.R. CP Conservation in the Presence of Instantons. *Phys. Rev. Lett.* **1977**, *38*, 1440–1443. [CrossRef]
6. Arvanitaki, A.; Dimopoulos, S.; Dubovsky, S.; Kaloper, N.; March-Russell, J. String Axiverse. *Phys. Rev.* **2010**, *D81*, 123530. [CrossRef]
7. Press, W.H.; Teukolsky, S.A. Floating Orbits, Superradiant Scattering and the Black-hole Bomb. *Nature* **1972**, *238*, 211–212. [CrossRef]
8. Brito, R.; Cardoso, V.; Pani, P. Superradiance. *Lect. Notes Phys.* **2015**, *906*, 1–237.
9. Arvanitaki, A.; Dubovsky, S. Exploring the String Axiverse with Precision Black Hole Physics. *Phys. Rev.* **2011**, *D83*, 044026. [CrossRef]
10. Herdeiro, C.A.R.; Radu, E. Kerr black holes with scalar hair. *Phys. Rev. Lett.* **2014**, *112*, 221101. [CrossRef]
11. Herdeiro, C.; Radu, E. Ergosurfaces for Kerr black holes with scalar hair. *Phys. Rev.* **2014**, *D89*, 124018. [CrossRef]
12. Ganchev, B.; Santos, J.E. Scalar Hairy Black Holes in Four Dimensions are Unstable. *Phys. Rev. Lett.* **2018**, *120*, 171101. [CrossRef] [PubMed]
13. Degollado, J.C.; Herdeiro, C.A.R.; Radu, E. Effective stability against superradiance of Kerr black holes with synchronised hair. *Phys. Lett.* **2018**, *B781*, 651–655. [CrossRef]
14. Akiyama, K.; Alberdi, A.; Alef, W.; Asada, K.; Azulay, R.; Baczko, A.-K.; Ball, D.; Baloković, M.; Barrett, J.; Bintley, D.; et al. [The Event Horizon Telescope Collaboration]. First M87 Event Horizon Telescope Results. I. The Shadow of the Supermassive Black Hole. *Astrophys. J.* **2019**, *875*, L1.
15. Akiyama, K.; Alberdi, A.; Alef, W.; Asada, K.; Azulay, R.; Baczko, A.-K.; Ball, D.; Baloković, M.; Barrett, J.; Bintley, D.; et al. [The Event Horizon Telescope Collaboration]. First M87 Event Horizon Telescope Results. V. Physical Origin of the Asymmetric Ring. *Astrophys. J.* **2019**, *875*, L5.

16. Akiyama, K.; Alberdi, A.; Alef, W.; Asada, K.; Azulay, R.; Baczko, A.-K.; Ball, D.; Baloković, M.; Barrett, J.; Bintley, D.; et al. [The Event Horizon Telescope Collaboration]. First M87 Event Horizon Telescope Results. VI. The Shadow and Mass of the Central Black Hole. *Astrophys. J.* **2019**, *875*, L6.
17. Cunha, P.V.P.; Herdeiro, C.A.R.; Radu, E.; Runarsson, H.F. Shadows of Kerr black holes with scalar hair. *Phys. Rev. Lett.* **2015**, *115*, 211102. [CrossRef]
18. Dolan, S.R. Instability of the massive Klein-Gordon field on the Kerr spacetime. *Phys. Rev.* **2007**, *D76*, 084001. [CrossRef]
19. Gebhardt, K.; Adams, J.; Richstone, D.; Lauer, T.R.; Faber, S.M.; Gultekin, K.; Murphy, J.; Tremaine, S. The Black-Hole Mass in M87 from Gemini/NIFS Adaptive Optics Observations. *Astrophys. J.* **2011**, *729*, 119. [CrossRef]
20. Walsh, J.L.; Barth, A.J.; Ho, L.C.; Sarzi, M. The M87 Black Hole Mass from Gas-dynamical Models of Space Telescope Imaging Spectrograph Observations. *Astrophys. J.* **2013**, *770*, 86. [CrossRef]
21. Nokhrina, E.E.; Gurvits, L.I.; Beskin, V.S.; Nakamura, M.; Asada, K.; Hada, K. M87 black hole mass and spin estimate through the position of the jet boundary shape break. *Mon. Not. Roy. Astron. Soc.* **2019**, *489*, 1197–1205. [CrossRef]
22. Tamburini, F.; Thidé, B.; Valle, M.D. Measurement of the spin of the M87 black hole from its observed twisted light. *arXiv* **2019**, arXiv:1904.07923.
23. Detweiler, S.L. Klein-Gordon equation and rotating black holes. *Phys. Rev.* **1980**, *D22*, 2323–2326. [CrossRef]
24. Zouros, T.J.M.; Eardley, D.M. Instabilities of massive scalar perturbations of a rotating black hole. *Annals Phys.* **1979**, *118*, 139–155. [CrossRef]
25. Volonteri, M.; Haardt, F.; Madau, P. The Assembly and merging history of supermassive black holes in hierarchical models of galaxy formation. *Astrophys. J.* **2003**, *582*, 559–573. [CrossRef]
26. East, W.E.; Pretorius, F. Superradiant Instability and Backreaction of Massive Vector Fields around Kerr Black Holes. *Phys. Rev. Lett.* **2017**, *119*, 041101. [CrossRef]
27. Herdeiro, C.A.R.; Radu, E. Dynamical Formation of Kerr Black Holes with Synchronized Hair: An Analytic Model. *Phys. Rev. Lett.* **2017**, *119*, 261101. [CrossRef]
28. Herdeiro, C.; Radu, E.; Runarsson, H. Kerr black holes with Proca hair. *Class. Quant. Grav.* **2016**, *33*, 154001. [CrossRef]
29. Brito, R.; Cardoso, V.; Pani, P. Black holes as particle detectors: evolution of superradiant instabilities. *Class. Quant. Grav.* **2015**, *32*, 134001. [CrossRef]
30. Herdeiro, C.; Radu, E. Construction and physical properties of Kerr black holes with scalar hair. *Class. Quant. Grav.* **2015**, *32*, 144001. [CrossRef]
31. Cunha, P.V.P.; Grover, J.; Herdeiro, C.; Radu, E.; Runarsson, H.; Wittig, A. Chaotic lensing around boson stars and Kerr black holes with scalar hair. *Phys. Rev.* **2016**, *D94*, 104023. [CrossRef]
32. Vincent, F.H.; Gourgoulhon, E.; Herdeiro, C.; Radu, E. Astrophysical imaging of Kerr black holes with scalar hair. *Phys. Rev.* **2016**, *D94*, 084045. [CrossRef]
33. Delgado, J.F.M.; Herdeiro, C.A.R.; Radu, E. Violations of the Kerr and Reissner-Nordström bounds: Horizon versus asymptotic quantities. *Phys. Rev.* **2016**, *D94*, 024006. [CrossRef]
34. Synge, J.L. The Escape of Photons from Gravitationally Intense Stars. *Mon. Not. Roy. Astron. Soc.* **1966**, *131*, 463–466. [CrossRef]
35. Bardeen, J.M. Timelike and Null Geodesies in the Kerr Metric. In *Black Holes*; Witt, C., Witt, B., Eds.; CRC Press: Boca Raton, FL, USA, 1973; p. 215.
36. Luminet, J.P. Image of a spherical black hole with thin accretion disk. *Astron. Astrophys.* **1979**, *75*, 228–235.
37. Luminet, J.P. An Illustrated History of Black Hole Imaging: Personal Recollections (1972–2002). *arXiv* **2019**, arXiv:1902.11196.
38. Falcke, H.; Melia, F.; Agol, E. Viewing the shadow of the black hole at the galactic center. *Astrophys. J.* **2000**, *528*, L13. [CrossRef]
39. Cunha, P.V.P.; Herdeiro, C.A.R. Shadows and strong gravitational lensing: A brief review. *Gen. Rel. Grav.* **2018**, *50*, 42. [CrossRef]
40. Cunha, P.V.P.; Herdeiro, C.A.R.; Radu, E.; Runarsson, H.F. Shadows of Kerr black holes with and without scalar hair. *Int. J. Mod. Phys.* **2016**, *D25*, 1641021. [CrossRef]
41. Grenzebach, A.; Perlick, V.; Lämmerzahl, C. Photon Regions and Shadows of Accelerated Black Holes. *Int. J. Mod. Phys.* **2015**, *D24*, 1542024. [CrossRef]

42. Teo, E. Spherical photon orbits around a kerr black hole. *Gen. Relativ. Gravit.* **2003**, *35*, 1909–1926. [CrossRef]
43. Walker, R.C.; Hardee, P.E.; Davies, F.B.; Ly, C.; Junor, W. The Structure and Dynamics of the Subparsec Jet in M87 Based on 50 VLBA Observations over 17 Years at 43 GHz. *Astrophys. J.* **2018**, *855*, 128. [CrossRef]
44. Abbott, B.P.; Abbott, R.; Abbott, T.D.; Abernathy, M.R.; Acernese, F.; Ackley, K.; Adams, C.; Adams, T.; Addesso, P.; Adhikari, R.X.; et al. [LIGO Scientific Collaboration and Virgo Collaboration]. Observation of Gravitational Waves from a Binary Black Hole Merger. *Phys. Rev. Lett.* **2016**, *116*, 061102. [CrossRef] [PubMed]
45. Abbott, B.P.; Abbott, R.; Abbott, T.D.; Abraham, S.; Acernese, F.; Ackley, K.; Adams, C.; Adhikari, R.X.; Adya, V.B.; Affeldt, C.; et al. [LIGO Scientific Collaboration and Virgo Collaboration]. GWTC-1: A Gravitational-Wave Transient Catalog of Compact Binary Mergers Observed by LIGO and Virgo during the First and Second Observing Runs. *Phys. Rev.* **2019**, *X9*, 031040. [CrossRef]
46. Davoudiasl, H.; Denton, P.B. Ultralight Boson Dark Matter and Event Horizon Telescope Observations of M87. *Phys. Rev. Lett.* **2019**, *123*, 021102. [CrossRef]
47. Hui, L.; Kabat, D.; Li, X.; Santoni, L.; Wong, S.S.C. Black Hole Hair from Scalar Dark Matter. *J. Cosmol. Astropart. Phys.* **2019**, *1906*, 038. [CrossRef]
48. Chen, Y.; Shu, J.; Xue, X.; Yuan, Q.; Zhao, Y. Probing Axions with Event Horizon Telescope Polarimetric Measurements. *arXiv* **2019**, arXiv:1905.02213.
49. Bar, N.; Blum, K.; Lacroix, T.; Panci, P. Looking for ultralight dark matter near supermassive black holes. *J. Cosmol. Astropart. Phys.* **2019**, *1907*, 045. [CrossRef]
50. Roy, R.; Yajnik, U.A. Evolution of black hole shadow in the presence of ultralight bosons. *arXiv* **2019**, arXiv:1906.03190.
51. Bambi, C.; Freese, K.; Vagnozzi, S.; Visinelli, L. Testing the rotational nature of the supermassive object M87* from the circularity and size of its first image. *Phys. Rev.* **2019**, *D100*, 044057. [CrossRef]
52. Tian, S.X.; Zhu, Z.-H. Testing the Schwarzschild metric in a strong field region with the Event Horizon Telescope. *Phys. Rev.* **2019**, *D100*, 064011. [CrossRef]
53. Vagnozzi, S.; Visinelli, L. Hunting for extra dimensions in the shadow of M87*. *Phys. Rev.* **2019**, *D100*, 024020. [CrossRef]
54. Contreras, E.; Rincón, A.; Panotopoulos, G.; Bargueño, P.; Koch, B. Black hole shadow of a rotating scale–dependent black hole. *arXiv* **2019**, arXiv:1906.06990.
55. Clough, K.; Ferreira, P.G.; Lagos, M. Growth of massive scalar hair around a Schwarzschild black hole. *Phys. Rev.* **2019**, *D100*, 063014. [CrossRef]
56. Narayan, R.; Johnson, M.D.; Gammie, C.F. The Shadow of a Spherically Accreting Black Hole. *Astrophys. J.* **2019**, *885*, L33. [CrossRef]
57. Gralla, S.E.; Holz, D.E.; Wald, R.M. Black Hole Shadows, Photon Rings, and Lensing Rings. *Phys. Rev.* **2019**, *D100*, 024018. [CrossRef]

© 2019 by the authors. Licensee MDPI, Basel, Switzerland. This article is an open access article distributed under the terms and conditions of the Creative Commons Attribution (CC BY) license (http://creativecommons.org/licenses/by/4.0/).

MDPI
St. Alban-Anlage 66
4052 Basel
Switzerland
Tel. +41 61 683 77 34
Fax +41 61 302 89 18
www.mdpi.com

Universe Editorial Office
E-mail: universe@mdpi.com
www.mdpi.com/journal/universe

www.ingramcontent.com/pod-product-compliance
Lightning Source LLC
LaVergne TN
LVHW070553100526
838202LV00012B/457